SOLID-STATE CHEMISTRY

A Modern Approach

SOLID-STATE CHEMISTRY

A Modern Approach

Ashok Kumar Jha, PhD

AAP | APPLE ACADEMIC PRESS

First edition published 2023

Apple Academic Press Inc.
1265 Goldenrod Circle, NE,
Palm Bay, FL 32905 USA

760 Laurentian Drive, Unit 19,
Burlington, ON L7N 0A4, CANADA

CRC Press
6000 Broken Sound Parkway NW,
Suite 300, Boca Raton, FL 33487-2742 USA

4 Park Square, Milton Park,
Abingdon, Oxon, OX14 4RN UK

Library and Archives Canada Cataloguing in Publication

Title: Solid-state chemistry : a modern approach / Ashok Kumar Jha, PhD.
Names: Jha, Ashok Kumar (Lecturer in chemistry), author.
Description: First edition. | Includes bibliographical references and index.
Identifiers: Canadiana (print) 20230146546 | Canadiana (ebook) 20230146570 | ISBN 9781774911976 (hardcover) | ISBN 9781774911983 (softcover) | ISBN 9781003328629 (ebook)
Subjects: LCSH: Solid state chemistry.
Classification: LCC QD478 .J43 2023 | DDC 541/.0421—dc23

Library of Congress Cataloging-in-Publication Data

CIP data on file with US Library of Congress

ISBN: 978-1-77491-197-6 (hbk)
ISBN: 978-1-77491-198-3 (pbk)
ISBN: 978-1-00332-862-9 (ebk)

About the Author

Ashok Kumar Jha, PhD, is Associate Professor in the University Department of Chemistry at T. M. Bhagalpur University, Bhagalpur, Bihar, India. Dr. Jha has a brilliant academic career to date. He has organized international and national seminars and webinars sponsored by the University Grants Commission (UCG), New Delhi; Council of Scientific and Industrial Research, New Delhi; Bihar Council on Science and Technology; National Bank for Agriculture and Rural Development; and other agencies. He has delivered many invited talks in India and abroad. Dr. Jha has teaching and research experience of 36 years. His interest is reflected through his many papers published in various national and international journals related to kinetics, coordination, and adsorption of arsenic and heavy metals. He has nine books to his credit. He has completed two research projects sponsored by UGC and has been working on the problem of arsenic and heavy-metal contamination and their removal. He teaches solid-state chemistry and quantum chemistry in postgraduate courses. Dr. Jha is a member of several academic societies and a member of editorial teams of many journals. He is currently the Vice-President of the Indian Chemical Society, Kolkata. He also publishes a biannual journal, *Jalchintan.*

Contents

Abbreviations

AFM	atomic force microscope
DSC	differential scanning calorimetry
DTA	differential thermal analysis
FTIR	Fourier transform infrared spectroscopy
HTSC	high-temperature superconductors
SEM	scanning electron microscopy
STM	scanning tunneling microscopy
TEM	transmission electron microscopy
TGA	thermogravimetric analysis
TMTSF	tetramethyl tetra selena fulvalene

Preface

This volume is a modern approach to solid-state chemistry that has been designed to cater to the needs of students, faculty, and researchers. I have made best possible effort to incorporate recent advances in solid-state chemistry, the latest techniques of characterization of solids, and several new dimensions of solid-state problems. Solid-state reactions, their kinetics, phase transitions, and magnetic properties have been discussed in detail. The electrical properties, such as semiconductors, metals, and insulators, have been described with suitable examples and proper explanations. Dielectric, piezoelectric, and ferroelectric properties of solids have been taken into account.

There is not an iota of doubt that solid surfaces act as catalyst and adsorption. Different theories of catalysis and adsorption isotherms have been dealt citing some experimental findings.

The contribution of chemists toward the developments of solid-state science is worth mentioning because they have developed novel methods of synthesis and hence made solid-state chemistry easy to understand. Though mathematical operations in a point group and space groups are not easy to understand, chemists have made the subject easier with the help of suitable molecules. Defects in crystals, irregularities on the solid surface, grain boundaries, steps, and kinks have been explained in the book, citing some good examples.

The different theories of superconductors, their properties, and application in development have also been incorporated. Intercalation chemistry with references to intercalation in layer structure has opened up a vast scope of research in catalysis and adsorption. Organic solids are, in general, not conductors, but a variety of organic solids show semiconductor and superconductor behavior. So, organic solids have been discussed with references to their preparation, properties, and structure.

A few solids such as clay and clay minerals have been characterized in my laboratory by employing the instrumental techniques of FTIR, PXRD, EDAX, and SEM. Those experimental techniques have been discussed in the book along with experimental findings. Bentonite samples have been characterized and studied for adsorption of toxic heavy metals from aqueous medium. Different types of adsorption isotherm and calculation of parameters have also been discussed.

The latest references in the book related to solid-state have been cited for further study for researchers opting to advance research in solid-state chemistry. Utmost care has been taken to include recent developments regarding the topic. But it is quite possible that a few points have slipped my mind, which will be included in the next edition along with suggestions from readers. This book in the present form will sure be a support for students, faculty, and researchers as the subject has been explained in the book in a very simple and lucid manner. I do hope that this book will be widely appreciated by the galaxy of solid-state researchers, students, and faculty worldwide.

Acknowledgments

This present book *Solid-State Chemistry: A Modern Approach* is based on recent advances in the field of solid-state chemistry. I have spent almost three years of my valuable time in completing this arduous task. It would be proper here to mention Prof. T. N. Gururow of I.I.S.C. Bangalore who helped me a lot and provided me with the facility of PXRD in the laboratory. Later on, I used the facility at the Central Institute of Fuel and Mines Research, Dhanbad, where principal scientist Dr. K. K. Mishra encouraged me to write a book on solid-state chemistry. The focus of my research was on solid-state and surface chemistry with a view to remediate metals through ion exchange and adsorption, so several experimental findings have been incorporated in the book. Thanks are also due to Prof. Bhabatosh Mondal, Department of Chemistry, Visva-Bharati University, Santiniketan, who acted like a catalyst in urging me to write a book on solid-state.

I would like to extend my sincere gratitude to Prof. D. C Mukherjee, former president Indian Chemical Society, Kolkata, who ignited my mind for novel approaches in research and book writing. I owe a great debt to him for his constant encouragement and motivation, and for revealing actual sense of the academic journey. Really, I take pride in meeting him and for his suggestions from time to time.

I would specially thank my research scholar, Sri Sachin Verma, alumni of I.I.T Kanpur for drawing the diagrams and figures of the book and other useful assistance. I would like to express my thanks to Dr. Ujjwal, another Ph.D. student of mine for the suggestion of publishing the book by Apple Academic Press.

I heartily acknowledge several organizations for the inclusion of technical questions prepared for different competitive examinations in India. The questions are from TIFR (Tata Institute of Fundamental Research), NET (National Testing Agency of the University Grants Commission), GATE (Graduate Aptitude Test in Engineering) and JNU (Jain Entrance Test (JET) of Jain of Jain University.

Help from various journals and published research papers has also been taken to present this book in the most lucid manner. Special care has been taken to prepare the manuscript with simple language. In addition, deep meditation and yoga also instilled confidence in me for delivering the best.

I am also thankful to my wife, Smt. Jyotsna Jha, for keeping me away from the home affairs during my journey of teaching and research.

I express my gratitude to all my teachers who helped to unearth the talent and potential lying unused within me during my journey of study from the beginning of my childhood. The grace of my teachers will never cease as it is eternal. I express my deep faith in Lord Siva for burning the light of knowledge within me and for inspiring me to take up the task of writing a book.

I also express my deep regard to my mother, Smt. Sudha Devi, whose desire to see me as a good academician and researcher promoted me to complete this project. Last but not the least, I express my thanks to all the fellows, students, and staff of the Department of Chemistry at T. M. Bhagalpur University who directly or indirectly helped me in this work.

Introduction

Solid-state reactions have become important in view of the preparation of solids, crystal growth, and single crystal of elements. The various methods of preparation of crystal growth have been discussed in this volume along with the kinetics of solid-state reactions. These studies help to explain structural properties of new materials. Preparation of new solids by taking advantage of recent experimental techniques has also been included in the text.

The desired structure and properties can also be studied by the crystal growth of a particular element.

To understand the characterization of solids, it is necessary to know the structural behavior, defects on the surface, and adsorption properties. PXRD is used to identify the diffraction pattern from different planes. Fourier transform infrared spectroscopy indicates the presence of functional groups through absorption peaks. SEM and TEM describe the surface properties of solids, which help analyze adsorption behavior on the surface. All these high-intensity electron be ams are used for electron penetration. X-ray absorption spectroscopy has also been used for structural information. Thermogravimetric analysis (TGA), differential thermal analysis (DTA), and differential scanning calorimetry (DSC) are the techniques employed for characterization of solids. During change of temperature, weight loss takes place, which indicates the presence of a particular solid. Exothermic and endothermic peaks in the graph also predict the nature of a solid. A weight loss of around 14% shows the presence of montmorillonite minerals. Curves depend on the structure and composition of solid. These studies also help in the determination of phase diagrams and transitions.

The knowledge of points groups and space groups are important to understand the coordinates in the crystal structure, rotation axis, translation axis, and plane of symmetry in the seven crystal systems. The unit cells of Bravais lattices and element of point and space symmetry have been discussed with a view to have a crystal clear three-dimensional image in mind.

The stoichiometric and nonstoichiometric defects in the crystal and grain boundaries in a crystal have become useful in understanding the solid surface behavior. Defects may be small or large depending on the temperature and other thermodynamic parameters. Kinetics of Frenkel and Shottky defects have been discussed along with the kinetics and equilibria. Solid do not attain perfectness even at ordinary temperatures and as a result dislocations, plane defects, and extended defects occur.

Solid-state chemistry has attained a new dimension due to the study of semiconductors and superconductors as well as applications in various fields. The main theme of Chapter 7 is the explanation of electronic properties by band theory. Explanation of conduction by metals, insulators, and semiconductors has been made on the basis of band theory. Physical as well as wavelength treatment of band theory have also been dealt with. K space, Brillouin zones, intrinsic and extrinsic semiconductors, doping with elements, and exhaustion range during temperature dependence and concentration have also been described in Chapter 9. Donor level and acceptor level in the bands of semiconductor has been shown by a clear diagram.

Superconductivity, the Meissner effect, isotope effect, and important characteristics of a superconductor have been explained in Chapter 5. Explanation of superconductivity on the basis of BCS theory has been included in the text. Examples of high-temperature superconductors (HTSC) of mixed transition metal and lanthanum nonstoichiometric oxides have been discussed. Various applications of superconductors in different fields of applied science are becoming very popular.

A few solids exist in two polymorphic forms. The transformation of one form to another is phase transition. The phase rule and condensed phase rule have been explained on the basis of thermodynamics considerations. Thermodynamic and kinetic factors both are important subjects to change in phases and so different phase diagrams of one and two components systems in solid solutions have been explained in the text. Triple point at which the phases are nonvariant explain the behavior of phases in water or sulphur system.

Application of diagrams of stable and metastable phases during phase transition provides a pathway of transition. Kinetics of phase transition from polymorph I and II as a function of temperature has been explained by the Arrhenius equation. Critical size of nuclei, nucleation rate, and Avrami equation explain the mechanism of phase transformation.

Martensitic transformations and order-disorder transitions are important in alloy system and disordered polymorphs.

Various transition metals and their oxides behave as heterogeneous catalysts due to physical and chemical binding of the substrate on the surface. This field of solid-state chemistry has opened up vistas of research in catalysis chemistry and adsorption. Empirical equations for adsorption, for example, Freundlich, Langmuir, and BET isotherms explain the adsorption behavior. Freundlich refers to multilayer adsorption whereas Langmuir explains monolayer adsorption. Thermodynamics of adsorption phenomena and spontaneity of the process have also been discussed. Physical adsorption is reversible whereas chemisorption is irreversible and requires energy of activation. Surface structure, that is, the presence of kink, terrace, and steps, is known by LEED method. Surface structure becomes important to understand the adsorption behavior. Surface area is known from the formula derived from the BET isotherm. Basic types of isotherms have also been explained by diagrams. Recent models of equations of isotherms have been introduced to explain the adsorption behavior of different adsorbents and adsorbates.

A glass is an amorphous solid long-range order that is prepared by cooling of the liquid state. Important oxides for glass formation are SiO_2, B_2O_3, and P_2O_5. Viscosity is an important property in the formation of glasses. So behavior of liquids on cooling has been studied.

Silicate glasses have generally two oxides, for example, $Na_2Si_2O_5$. B_2O_3 and SiO_2 are important constituents of silicate glasses.

CHAPTER 1

Solid-State Reactions

ABSTRACT

Several methods of preparation of solid-state reactions are available. Phase boundary reactions of spinel oxides have been described. Wagner mechanism has explained the mechanism of the reaction between MgO and Al_2O_3. Surface area of solids, reactivity of solids and surface structure are the factors affecting the rate of solid-state reactions. There are limitations of solid-state reactions. Intercalation compounds of transition metal dichalcogenides with metal cations with suitable examples give an insight into the mechanism of solid-state reactions. Growth of single crystals, epitaxial growth of thin layers and zone melting are few possible mechanisms of methods of preparation.

1.1 INTRODUCTION

Alhough solid-state reactions are not very common reactions, the reactions between oxides of metals halides and carbonate were studied by Hedvall (1955).

For solid-state reactions, thermodynamic and kinetic aspects have to be considered.

The general solid-state reaction mechanism involves the initial formation of one or more solid products that separate the reactions.

Solid-state reactions are exothermic and the free energies change the products and the reactants decide the course of the reaction. This requires the chemical potential of the components.

The problems of solid-state reactions can be induced by the reaction of two crystals of MgO and Al_2O_3. After heating a layer of $MgAl_2O_4$ is

Solid-State Chemistry: A Modern Approach. Ashok Kumar Jha, PhD (Auth.)

formed at the interface. The formation of $MgAl_2O_4$ nuclei is the nucleation caused by the reorganization of reactants and products.

Ions such as Mg^{2+} in MgO and Al^{3+} in Al_2O_3 are regarded as being trapped on their lattice sites. At very high temperatures, such ions have sufficient thermal energy to leave their own site and through the crystal.

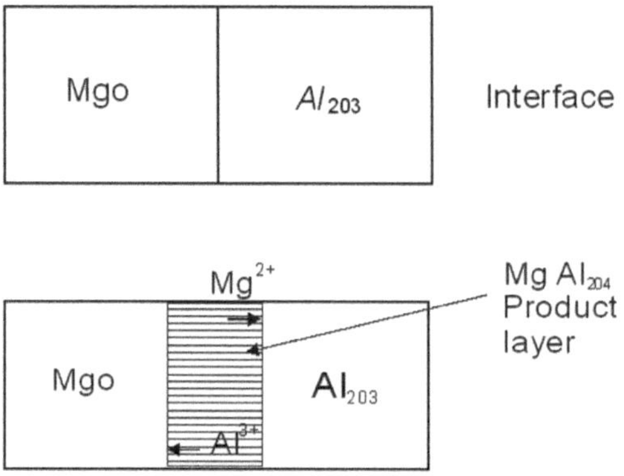

FIGURE 1.1 Interdiffusion of cations, of single crystals of MgO and Al_2O_3 to give $MgAl_2O_4$.

The diffusion of Mg^{2+} ions must occur right through the existing $MgAl_2O_4$ product layer. Two reaction center faces are possible as follows:

i) The reaction between MgO and $MgAl_2O_4$
ii) The reaction between $MgAl_2O_4$ and Al_2O_3

The rate-determining step for the further reaction is the diffusion of Mg^{2+} and Al^{3+} ions to and from these interfaces.

1.2 SPINEL OXIDES

The formation of spinel oxides can be generally set as AO/AB_2O_4. The phase boundary reaction at the AO/AB_2O_4 interface consists of homogenous occupancy of cations in the octahedral and tetrahedral sites of the CCP structure.

It has been observed that in the reaction between single-crystalline MgO and polycrystalline Cr_2O_3, the product forming at the MgO/Cr_2O_3 interface is also monocrystalline. Similarly, in the reactions between polycrystalline NiO and single-crystalline Al_2O_3, the phase boundary is also monocrystalline.

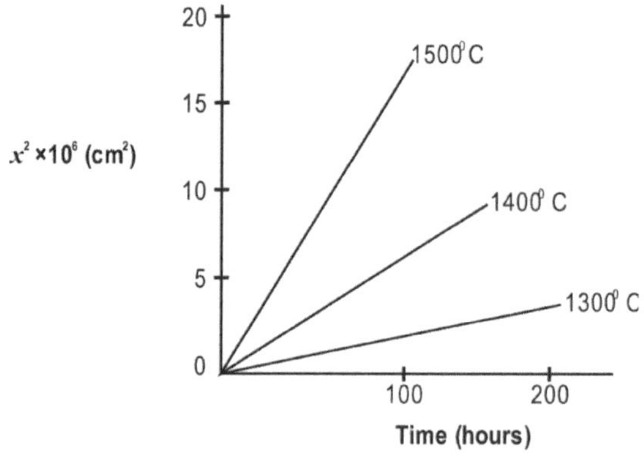

FIGURE 1.2 Thickness vs time.

Thickness (x) of $NiAl_2O_4$ product layer as a function of temperature and time formed by the reaction of NiO and Al_2O_3 (Pettit et al., 1966).

1.3 REACTION RATES

The reaction between NiO and Al_2O_3 fallow a parabolic-rate law when the thickness of the product layer exceeds 1 min. The parabolic rate has been calculated with the help of self-diffusion coefficients of captions through the spinal and the Gibbs free energy formation of the product. Let x be the amount of reaction (thickness of the spinel layer growing); t be the time; and k and k' be the rate constants. Therefore, according to parabolic rate law, we write the reaction rate as

$$\frac{dx}{dt} = k'x - 1$$

$$x = (k't)^{\frac{1}{2}}$$

Results are shown for different temperatures in Figure 1.2; x^2 is plotted against time and straight-line plots are obtained of slopes equal to k'. The reaction occurs fast with increasing temperature; the activation energy for the reaction is obtained from a plot of logk' vs 1/T.

1.4 WAGNER REACTION MECHANISM

Wagner mechanism has explained the mechanism of the reaction between MgO and Al_2O_3 involving the counter diffusion of Mg^{2+} and Al^{3+} ions through the product layer for every three Mg^{2+} ions diffusing to the right-hand interface, two Al^{3+} ions must diffuse to the left-hand interface. The reactions occurring at the two interfaces are as follows:

i) Interface MgO/Mg Al_2O_4

$$2Al^{3+} + 3Mg^{2+} + 4MgO \rightarrow Mg\,Al_2O_4$$

ii) Interface Mg Al_2O_4/Al_2O_3

$$3\,Mg^{2+} + 2\,Al^{3+} + 4Al_2O_3 \rightarrow 3Mg\,Al_2O_4$$

Overall reaction, $4\,MgO + 4\,Al_2O_3 \rightarrow 3MgAl_2O_4$, gives three times as that of the spinel product provided in reaction (i). Hence, the right-hand interface should grow or move at three times the rate of left-hand interface.

Inert markers have been used to know the mechanism of spinel formation. Before the start of the reaction, a thin platinum (Pt) wire is placed at the interface. When the reaction proceeds, the position of the marker prompts about the mechanism of diffusion. The pores used as natural markers support the counter diffusion of captions in oxide spinal formation reactions.

In reactions, in which the interfaces between reactant and product can be seen due to differences in color, the movement of the interface is used as a marker to monitor the progress marker known as the Kirkendall effect.

1.5 FACTORS AFFECTING THE RATE

1.5.1 SURFACE AREA OF SOLIDS

The surface area of a given amount of solid varies if present as a fine powder.

The surface area of reacting solids affects the reaction rates due to the toolbar area of contact between the grains of the reacting solids.

Although the surface area of solids controls the area of contact between reacting grains in a mixture, it does not appear in the equation for the rate of reaction. The thickness of the product layer x decreases with decreasing particle size. Thus, particle size and surface area affect the value of x.

1.5.2 REACTIVITY OF SOLIDS

Nucleation of the product is facilitated if there is a structural similarity between the product and one or both of the reactants. In the reaction between MgO and Al_2O_3, the spinel product has a similar oxide ion arrangement to that in MgO: Spinel nuclei may form at the surface, and the nucleating spinel makes use of matching of its own structure to that of an existing phase (MgO) in the reaction mixture.

1.5.3 SURFACE STRUCTURE

The nucleation varies with the surface structure of the reacting phases. It is not necessary that the structure should be the same.

Since different surfaces have different structures, their reactivity differs considerably.

If NaCl crystallites from an aqueous solution, the normal morphology is as cubes presenting [100] faces.

On addition of urea in the solution, octahedral crystals are formed having [111] face.

Urea influences the reactivity of various surfaces.

Those surfaces that are most prominent in a crystal are those that grow most slowly.

It has been established now that the different parts of the surface of a crystal have different structures and reactivity. The reactivity of solids also depends on the presence of crystal defects. The original surface of the nucleus is not exactly planar but convenient sites are generated for atoms, ions, or molecules to attach themselves and as a result, the growth of the crystals became faster.

The equation $x = (k't)^{\frac{1}{2}}$ expresses the rate of solid-state reaction for the diffusion of ions through the bulk of the crystals. The diffusion of ions is enhanced by the presence of crystal defects. The defects may be intrinsic to the crystals or by the presence of impurities.

1.6 DIFFICULTIES

The study of solid-state reactions is difficult (1) at the temperature of 1500°C spinel existed as a range of solid solutions on the join $MgO–Al_2O_3$.

The solid solution compositions cover the range 50–60 mol% Al_2O_3, in the initial stages of the reaction, the spinel product may have a variable composition. The spinel produced at the MgO-spinel interface will be the most magnesium-rich spinel, $MgAl_2O_4$, but at the Al_2O_3–spinel interface will be the most magnesium-deficient spinel, $Mg_{0.73}Al_{2.18}O_4$. (2) During cooling of partially reacted crystals, the spinel product breaks away from the MgO parent crystal due to the difference in the volume of the phases as a result of which the study of solid-state diffusion becomes difficult.

1.7 EXPERIMENTAL

1.7.1 HEAT TREATMENT

Both MgO and Al_2O_3 are refracting materials that cannot react together below 1200–1300°C. The final product is obtained after heating for several hours. During heating, sintering and grain growth of both reactant and product phases occur as a result surface area is reduced. The effect of grinding is to maintain a high surface area. The reaction rate may also be speeded by making pallets prior to heating.

At high temperature, the reaction rates are accelerated. The volatilization of one or more components of the reacting mixture may take place at a high temperature. Alkali oxides volatilize at high temperatures. It is, therefore, not possible to give a safe temperature below which volatilization does not occur.

In the case of Li_2CO_3, melting begins to occur at 700°C. When mixed with other components, the evaporation rate is reduced. So, care should be taken to fix a suitable temperature for heating.

1.7.2 REAGENTS

Starting materials should be dried thoroughly prior to weighing. Fine-grained materials are used in order to maximize surface areas. Both MgO and Al_2O_3 are the starting materials for the synthesis of spinel $MgAl_2O_4$. Also, $MgCO_3$ could be used as the source of MgO since it is less hygroscopic than MgO.

It is not mandatory to facilitate decomposition of $MgCO_3$ separately but mixed with $Al_2O_3$for heating.

1.7.3 MIXING

The reactants are weighed in the required amount and mixed together. The volatile organic liquid is added for homogenization of the mixture. During the process of grinding and mixing, the organic liquid volatilizes in 10–15 min. Manual mixing is not preferred if the quantity is larger. So, mixing is done in batches. Mechanical mixing may be preferred by using a ball mill that may take several hours.

For the reactions at high temperatures, the container material is chemically inert to the reactants. Platinum and gold are the noble metals found suitable. The container may be crucibles or boats made from foils. Nickel may be used for low-temperature reactions, for example, 600–700°C. Refractory inorganic materials may be used for containers, for example, crucibles of Al_2O_3, stabilized ZrO_2, and SiO_2.

1.7.4 ANALYSIS

The main technique for analyzing the product is X-ray powder diffraction. Each crystalline material has its own characteristics X-ray powder pattern. This shows whether the reaction is complete or not. The X-ray powder method does not give a chemical analysis of the product. X-ray fluorescence of atomic absorption analysis may be used to determine the overall chemical composition.

1.8 CO-PRECIPITATION

Co-precipitation procedures can be adopted to achieve homogenization together with small particle size. Oxalates of zinc and iron are used as the reactants in the synthesis of $ZnFe_2O_4$ spinel. The oxalates of zinc and iron are dissolved in water in a 1:1 ratio, mixed, and heated to evaporation. Oxalates of zinc and iron are gradually precipitated together, and hence fine powdered solid solution is obtained. The precipitated solids are filtered and calcined. A temperature of 1000°C is needed for the formation of $ZnFe_2O_4$. The reaction may be expressed as

$$Fe_2\left(\left(COO\right)_2\right)_3 + Zn\left(COO\right)_2 \rightarrow ZnFe_2O_4 + 4CO + 4CO_2$$

The method has been applied for the preparation of spinels, for example, $COFe_2O_4$, $MnFe_2O_4$, and $NiFe_2O_4$.

1.8.1 LIMITATIONS

(A) This method is not effective when the reactants used have different solubilities in water.
(B) Precipitation of the reactants does not take place at the same rate.

(ii) Ferrite spinels such as $NiFe_2O_4$ have been prepared from a basic double acetate pyridinate of nickel and iron $Ni_3Fe_6(CH_3COO)_{17}O_3OH \cdot 12C_5H_5N$. The Ni:Fe is 1:2 in this compound and purification is carried out by recrystallization from pyridine. The pyridinate is prepared from a basic double acetate hydrate phase. For $NiFe_2O_4$ crystals, the pyrididnate crystals are heated to 300°C in order to remove organic materials followed by heating in air at 1000°C for 2–3 days.

Chromite spinels $MnCr_2O_4$, $CuCr_2O_4$, etc. can be synthesized by this method. $MnCr_2O_7 \cdot 4C_5H_5N$ is heated gradually to 1500°C to give $MnCr_2O_4$ which is reduced to trivalent chromium. The mixture is heated in a hydrogen-rich atmosphere at 1100°C for maintaining the divalent state of manganese. These precursor methods are capable of yielding phases of accurate stoichiometry. However, this method is not applicable to all syntheses as each synthesis requires its own conditions.

1.9 KINETIC STUDY

The slowest step is the rate-determining step. So, it is desired to know the slowest step by considering the following possibilities:

a) Transport of matter to the reaction interface.
b) Reaction at the interface.
c) Transfer of matter away from the reaction interface.

The following equation represents the rate of concentration change:

$$\frac{dc}{dt} = -kc^n$$

where n represents reaction order and k represents constant. However, it may be often misleading as the reactions do not involve molecules. It is found that n is not a simple integer but may be fractional.

1.10 INTERCALATION CHEMISTRY

Intercalation means insertion of extra atoms or ions into the structure or by ion exchange, for example, replacement of ions in the structure with other ions. The structure should be such that foreign atoms or ions are allowed to diffuse readily. Intercalation may also be explained as the reversible insertion of guest species into the host structure. The host has a number of unoccupied sites. When the guest atoms or ions penetrate the space between the layers, the layers are pushed apart. If the intercalated atoms diffuse out of the crystal, the layers move back together. Synthetic or natural clay, for example, zeolite possesses three-dimensional aluminosilicate structures containing cages and tunnels. These contain hydrated cations which may be exchanged by other cations by ion exchange. Owing to ion exchange capacity, zeolites and naturally occurring clay minerals, for example, Bentonites have many industrial applications such as water softeners.

Graphite also acts as a host crystal in which a wide variety of atoms, ions, and molecules can be intercalated. It is possible to intercalate alkali cations, halide anions, amines, oxysalts, and metal halides between the carbon layers. Some examples of graphite-intercalated compounds are as follows:

$$\text{Graphite} \xrightarrow[250^{o}C]{HF/F_2} \text{graphite fluoride, } C_{3\cdot 6}F \text{ to } C_4F$$

(Black)

$$\text{Graphite graphite fluoride, } CF_{0.68} \text{ to } CF$$

(White)

$$\text{Graphite} + \text{K (melt or vapor)} \rightarrow C_8K \text{ (bronze)}$$

When alkali metal is inserted, the host structure expands along *c* direction. Alkali metal size decides the separation between the adjacent carbon planes. The increase in *c* parameter is 0.38 Å for Li, 1.15 Å for Na, and 2.60 Å for Cs. The graphite changes its magnetic behavior from diamagnetic to paramagnetic and as a result increase in conductivity takes place along the *c* direction due to electron transfer resulting from intercalation. Alkali intercalates of graphite becomes colored due to electron transfer and hydride is formed with hydrogen. Golden KC_8 is converted into blue $KH_{0.67}C_8$ in the presence of hydrogen.

Several electron acceptors such as HSO_4, NO_3, CrO_2Cl_2, CrO_3, MoO_3, $FeCl_3$, and $FeCl_2$ may be successfully intercalated into graphite where it acts as a donor of electrons. With the Hall effect measurement, it is clear that the electron transfer is of *p*-type carriers. The conductivities of such intercalated graphites are similar to metallic aluminum.

It has also been established that guest species intercalated into graphite are well ordered, giving rise to staging. A monolayer of guest molecular ions gets inserted between every pair of graphite layers in case of the highest concentration known as the first-stage compound. In a second-stage compound, every alternate layer gets occupied by the guest. In a third-stage compound, guest species get inserted between every third layer.

The interconversion of one stage to the next takes place.

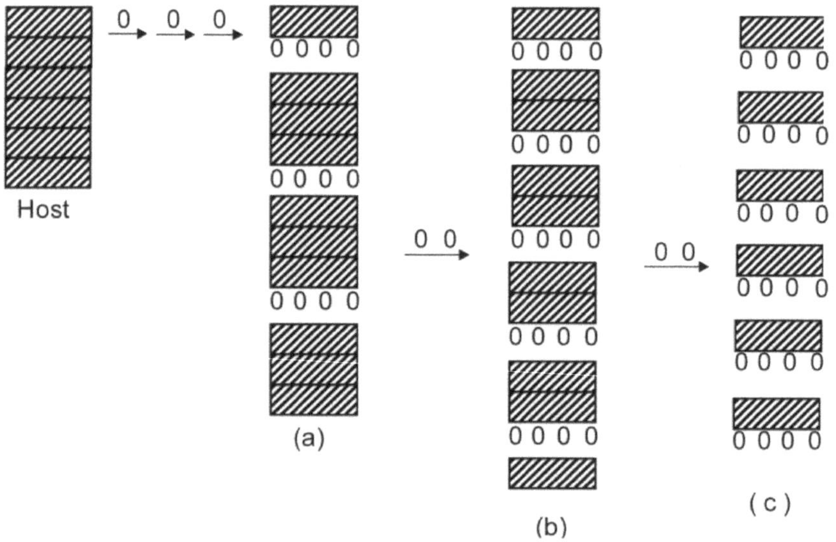

FIGURE 1.3 Schematic representation of staging in graphite intercalation (a) third stage (b) second stage (c) first stage.

1.10.1 *TRANSITION-METAL DICHALCOGENIDES*

Intercalation compounds of dichalcogenides are of considerable importance. Transition–metal dichalcogenides MX_2 of group IV, V, and VI (M = Ti, Zr, V, Nb, W, and so on.) have hexagonal or rhombohedral layered structures. The unit cell of the structure of these compounds is the

MX_2 sandwiching in which the transition metal M occupies octahedral sites between two layers of close-packed chalcogen X atoms. Stacking of X–M–X sandwiches takes place along the c-axis in the hexagon. Bonding within MX_2 unit is strong whereas bonding between different MX_2 units is weak Van der Waals. It is these weak Van der Waals gaps that are responsible for intercalation. Intercalation compounds of dichalcogenides may be divided into the following classes:

a) The Lewis base has a lone pair of electrons, for example, ammonia, *n*-alkyl amines, pyridines, etc.
b) Metal or molecular cations, Li, Na, K, or $[(C_2H_5)_2Co]^+$.
c) Compounds containing both cations and neutral polar molecules in the Van der Waals gas.

It has been established that disulfides and diselenides form stable intercalates. The intercalation results in an expansion of the lattice expands along the *c*-axis in intercalation. The nature of the host and guest molecules determine the stability and stoichiometry of intercalated compounds. Intercalation with organic compounds has been carried out by direct reaction of the chalcogenide with organic liquid.

One of the examples is the reaction of *n*-alkyl amines ($C_nH_{2n+1}NH_2$) with superconducting host $2HTaS_2$. The whole series of intercalates from $n = 1$ to $n = 18$ has been obtained by the direct reaction.

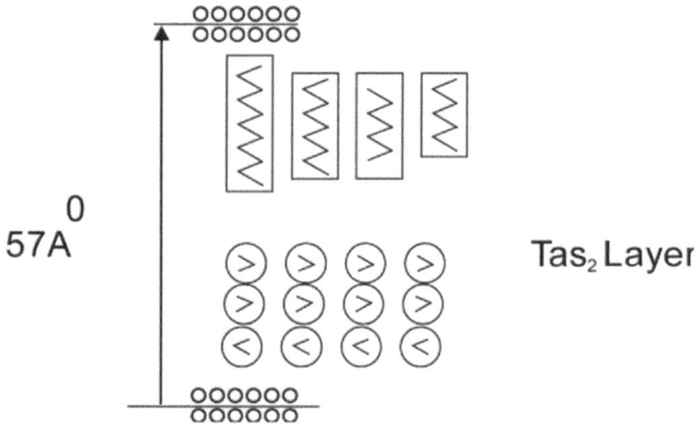

Structure of TaS_2–*n*-alkylamine (a) bilayer (b) monolayer configuration of the intercalated amine.

1.10.2 INTERCALATION COMPOUNDS OF TRANSITION METAL DICHALCOGENIDES WITH METAL CATIONS

The disulfides of transition metals have layered structures, for example, TiS_2. It has a sandwich structure in which titanium occupies the octahedral site between layers of sulfur atoms. A simple intercalation reaction is a reaction between lithium and TiS_2 to form Li_xTiS_2, where $0 < x < 1$. Lithium ions enter the space between TiS_2 sandwiches. The reaction is usually carried out in an organic solvent by using *n*-butyllithium as the source of Li.

$$x\, C_4H_9Li + TiS_2 \xrightarrow[inertatmosphere]{Hexane} Li_xTiS_2 + \frac{x}{2}\, C_8H_{18}$$

Li_xTiS_2 may be filtered and washed with hexane.

Schematic layered structure of (a) TiS_2 and (b) $LiTiS_2$ are shown as follows:

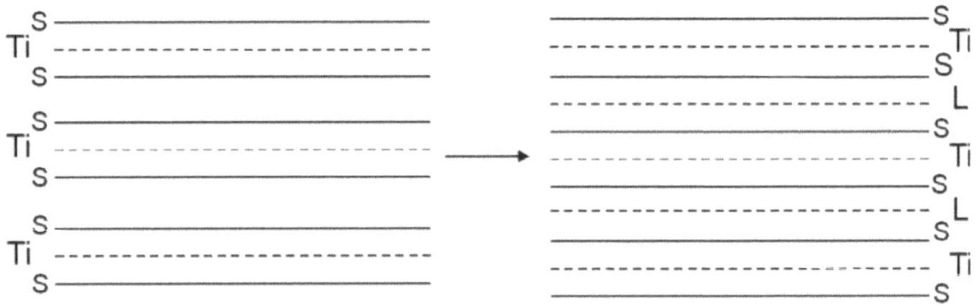

(a) Tis_2 (b) Li Tis_2

1.10.3 PREPARATION

Furthermore, Li_xTiS_2 may be prepared by electrochemical method. Also, TiS_2 powder bonded on to a Teflon polymer is dipped in $LiClO_4$ dissolved in a polar non aqueous solvent, dioxolane. It works as a cathode A sheet of Li metal dipped into the electrolyte serves as the anode. On completing the external circuit, an electron current flows in the external circuit and Li^+ ions are transported from the anode to the cathode through the electrolyte and intercalate the TiS_2.

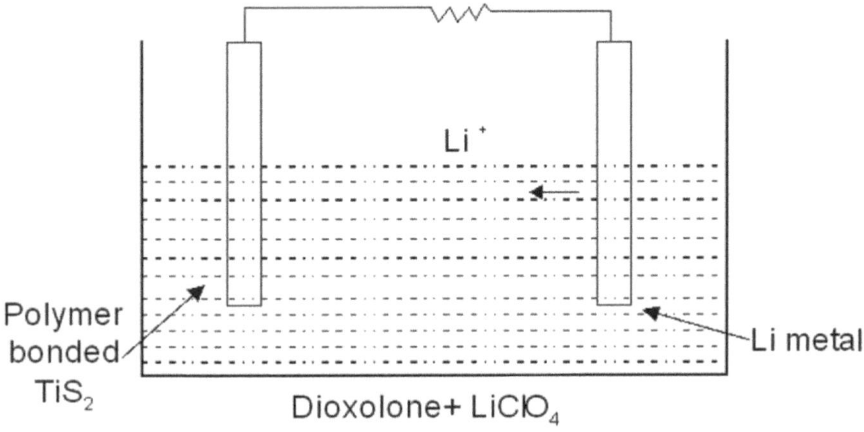

Cell reaction: $Li - Li^{+} + e^{-} \rightarrow Li_2$ Anode

$TiS_2 + Li^{+} + e^{-} \rightarrow LiTiS_2$ Cathode

1.10.4 ELECTROCHEMICAL INTERCALATION OF LITHIUM BY TiS_2

By electrochemical methods, intercalation compounds using other alkali cations, Cu^{+}, solvated ion, Ag^{+}, H^{+}, NH_3, and metallocenes, are possible. A large number of layered and tunnel compounds may form intercalated compounds.

1.10.5 PREPARATION OF THIN FILMS

Thin films are used to protect coatings on materials and so very important in industries Due to their very large surface area to volume, thin films have special properties.

1.10.6 CATHODIC DEPOSITION

In this method, two metal electrodes are dipped into an electrolye solution and metal ions after reduction get deposited in the form of thin film at cathode.

1.10.7 ELECTROLESS DEPOSITION

This takes place in the absence of an applied external field. It is used for the deposition of nickel films. The disadvantage of the deposition of metallic films is limited only onto substrates, for example, metals. This is not suitable for the electronically non conducting substrates.

1.10.8 ANODIC OXIDATION

This is an electrolytic method for producing oxide films on the surface of metals such as Al. Ta, Nb, Ti, and Zr. These metals act as anodes and are dipped into a liquid electrolyte. Oxide ions are attracted to the anode to form a thin layer, for example, Al_2O_3. If the field strength is increased, a sizable number of O_2^- ions diffuse to the metal surface. Thus, the oxide layer grows thicker.

1.10.9 THERMAL OXIDATION

Several substances get oxidized in air at high temperatures and form an inert film inhibiting further oxidation. Aluminum forms an oxide film at room temperature but the thickness increases with increasing temperatures. This is not restricted to oxide films.

1.10.10 CHEMICAL VAPOR DEPOSITION

This method is used to obtain purest films of semiconductors. Moreover, III–V semiconductor compounds can also be obtained by this method. The decomposition may be achieved by pyrolysis, photolysis, or chemical reaction

$$GeH_4 \xrightarrow{Heat/h\nu} Ge$$

$$Si(CH_2CH_3)_4 \xrightarrow[Air]{Heat} SiO_2$$

$$SiCl_4 + 2H_2 \rightarrow Si + 4HCl$$

$$SiH_4 \rightarrow Si + 2H_2$$

Vapor deposited films tend to match the structure of the substrates resulting in epitaxial growth mechanisms. These films may grow by crystallization or precipitation from a liquid phase into the substrate.

1.10.11 VACUUM EVAPORATION

This is the most important method used for the preparation of thin films. A high vacuum of 10^{-6} torr is required for operation of the system. When materials from the evaporation source is subjected to heating or electron bombardment, gaseous phase is obtained. The gaseous material then deposits on the substrate. Some of the substrates used are ceramics (Al_2O_3), glass, alkali halides, silicon, and germanium. Source materials for evaporation may be used such as metals, alloys, semiconductors, and insulators.

The substrate should be such that the films adhere to it in a good manner.

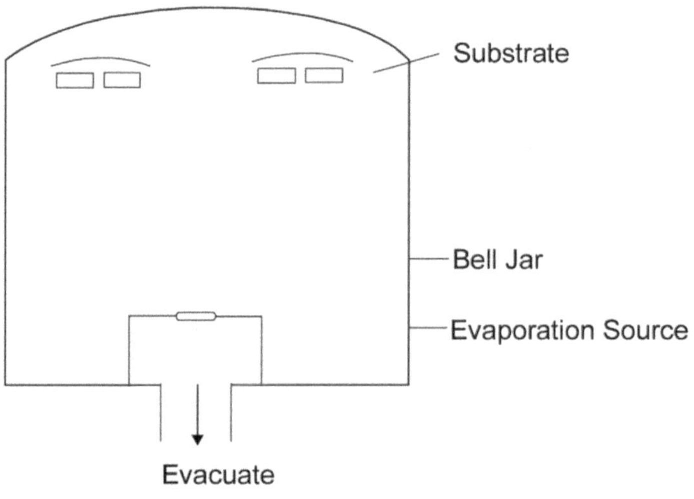

Vacuum evaporation equipment for thin film deposition.

1.11 GROWTH OF SINGLE CRYSTALS

The growth of crystals takes place from liquid vapor or solid phase. Only vapor and liquid give crystals of sufficient size to be used in applications.

1.11.1 CZOCHROLSKI METHOD

This method can be adopted for single crystal growth from melt. A single crystal is allowed to have contact with the melt. The temperature of the melt is maintained above its melting power. When the seed is drawn out of the melt, the melt changes to solid at the seed surface which results in a rode shape crystal. New developed crystal retains the crystallographic orientation as the original seed. The melt and growing crystal are rotated counter clock wise during pulling for maintaining a constant temperature. This method is used for the growth of semiconducting crystals such as Si, Ge, GaAs, etc. An inert gas atmosphere at high pressure is required to present, loss of As, P etc.

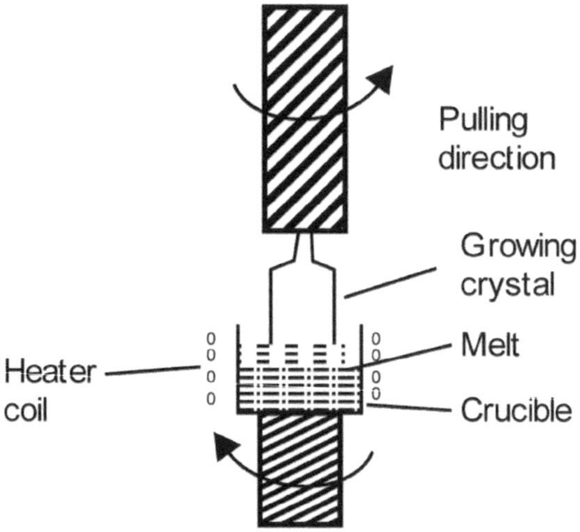

FIGURE 1.4 Czochralski method for crystal growth.

1.11.2 BRIDGMAN AND STOCKBARGER METHODS

In this method, the melt is solidified by when the melt is subjected to pass through a temperature gradient. The melt solidifies at the cooler end. Temperature gradient furnace is used for melt in the Bridgman method where solid phase separated at the cooler end. Use of seed crystal takes place in this method.

1.11.3 ZONE MELTING

This method is used for the purification and crystal growth of tungsten. The impurities usually concentrate in the liquid than in the solid phase. As the molten zone moves, pure solid solidifies and impurities are swept out.

1.11.4 FLUX METHOD OR PRECIPITATION METHOD

Crystals grow from a solvent of varying composition to crystals in the precipitation method. The solvent may belong to one of the constituents of the desired crystal. The solvent melts are referred to as fluxes. This method has been used to grow crystals of B alumina using a borate flux.

1.11.5 EPITAXIAL GROWTH OF THIN LAYERS

Epitaxial growth means two-dimensional structural relationship between seed and crystal. In the epitaxial growth of thin layers, oriented growth of the crystal occurs on the surface of a substrate. The substrate crystal may be the same or it may be different material whose lattice parameters at its surface match those of the growing crystal. Thin layers of GaAs have been grown from the vapor phase by deposition on the substrates such as Al_2O_3. Thin layers may also be prepared by epitaxial growth from liquids. Single crystals in the form of thin layers are required for application in electronic devices.

1.11.6 HYDROTHERMAL METHODS

In hydrothermal method, the water or its vapor serves as the pressure transmitting medium. This method is suitable for the synthesis of phases that are unstable at a higher temperature. It is also applied for the growth of single crystals. A suitable temperature gradient is in the reaction vessel where the dissolution of the starting material takes place at the hot end and precipitation at the cooler end.

KEYWORDS

- **solid-state reactions**
- **intercalation**
- **limitations**
- **transition-metal dichalcogenides**
- **spinel oxides**

CHAPTER 2

Characterization of Solids

ABSTRACT

Thermal analysis is done with a view to characterize the samples. The main thermal analysis techniques used are TGA, DTA and DSC. TGA is the weight change as a function of time whereas DTA refers to the temperature change as a function of temperature. DSC measures the enthalpy change as a function of temperature. TGA and DTA of bentonite characterizes the sample. Scanning electron microscopy studies the surface morphology of a material or adsorbent. Scanning Tunneling Microscopy is used for a view of the atomic structure and electron states. EDS or EDX technique is used for elemental analysis and their relative proportion. X-ray diffraction is used to know the diffraction pattern of a crystal and measurement of intensity gives a clue to test crystal structure. The bands in FTIR show the stretching frequency of the bonds.

2.1 INTRODUCTION

Differential thermal analysis (DTA) or differential scanning calorimetry (DSC) is employed to know phase transitions. These thermal decompositions are helpful in the determination of phase diagrams. Thermogravimetric analysis (TGA) records the weight loss of a sample as a function of temperature. The DTA measures the temperature difference between a sample and an inert reference material with increasing temperature. The DTA gives the value of changes in heat content. Differential scanning calorimetry measures enthalpy changes quantitatively in a sample with temperature. Nowadays, automatic thermal analysis equipments are available which can do all analyses done by TGA, DTA, and DSC.

Solid-State Chemistry: A Modern Approach. Ashok Kumar Jha, PhD (Auth.)

2.2 THERMOGRAVIMETRIC ANALYSIS (TGA)

Thermogravimetry is used to measure the change in weight of a sample with time which can be seen in the recorded graph. When a few milligrams of a sample is heated at a constant rate, it starts decomposing at temperature, *T*. Decomposition takes place over a range of temperature T_1 to T_2 showing first weight loss and second weight loss. Weight losses during heating are used for knowing compositional changes. The smectite group of minerals is characterized by TGA studies. The first weight loss during TGA shows a loss of hygroscopic water at around 2000°C and the second weight loss is explained due to dehydroxylation. Clay minerals are characterized by thermogravimetric analysis, for example, illite, kaolinite, vermiculite, and bentonite. Thus, a change in weight corresponds to dehydration of the sample which shows up on DTA as an endothermic event. If the DTA effect is exothermic, the sample is metastable which shows a decrease in enthalpy of the sample finally changing to a stable structure.

If we start with a hydrated material, dehydration takes place at first and then dehydroxylation. Dehydration of hygroscopic water takes place at a certain temperature around 100°C and the graph appears as an endotherm. Polymorphic transition takes place at some higher temperature. This process is also endothermic. When the sample melts, it gives a third endotherm. If a process on heating is endothermic, the reverse process on cooling is exothermic. When clay or clay minerals are heated, the process is endothermic. Endothermic peaks during heating of clay confirm presence of bentonite, kaolinite, illite, vermiculites, etc.

Heating of a crystalline substance is an endothermic process, for example, silica. When silica is heated, an endotherm appears on melting. On cooling, the liquid does not recrystallize but becomes supercooled. Thus, for SiO_2 the liquid is viscous even above the melting point of 1700°C.

2.3 PHASE TRANSITION

Polymorphic phase transitions are studied easily and accurately by DTA. The DTA provides monitor of both the property and composition of the material. If a material is completely unknown, DTA alone cannot characterize the material. TGA is employed to determine purity by comparing the weight loss for a particular material with the theoretical value of the pure material. If a montmorillonite unit is present in clay, the total weight loss is 14%. In some cases, the total weight loss is 7 or 9% indicating the presence of impurities in the pure sample. Thus, by employing TGA, we can characterize the material and determine the purity of the sample. First weight loss during TGA is due to loss of hygroscopic water and the second weight loss is due to dehydroxylation.

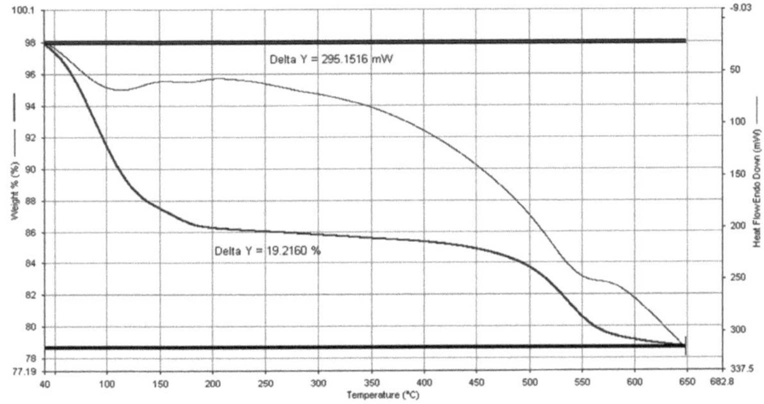

FIGURE 2.1 TGA and DTA of Rajmahal bentonite mineral S_1.

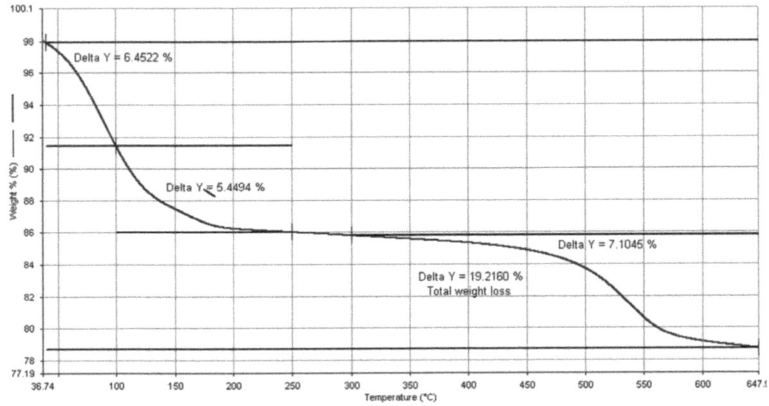

FIGURE 2.2 TGA and DTA of a sample of Rajmahal bentonite mineral S_2.

Total weight loss here in the sample is 19.2160% which is due to montmorillonite mineral.

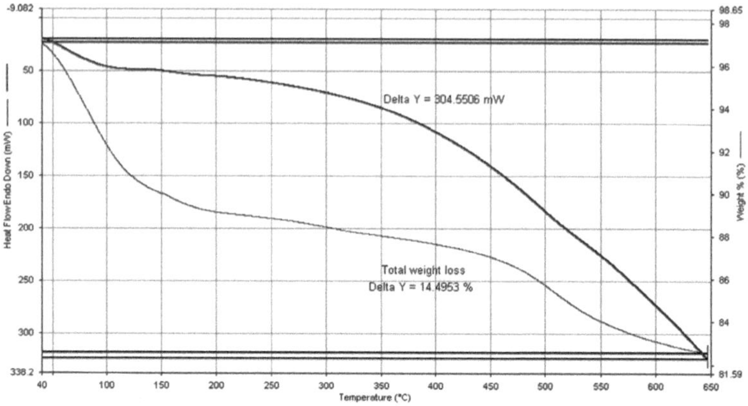

FIGURE 2.3 TGA of bentonite sample S_3.

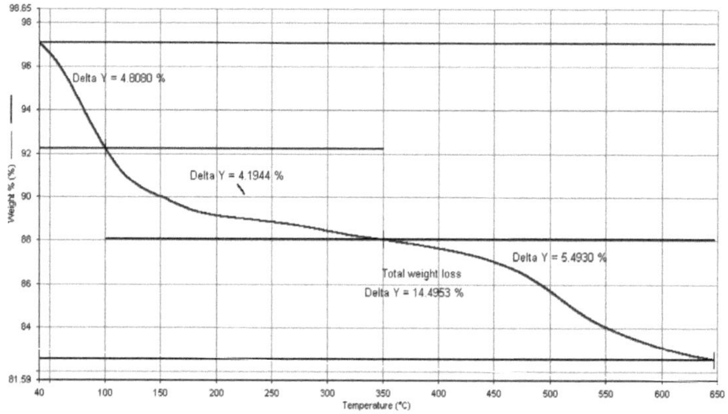

FIGURE 2.4 TGA of sample S_4.

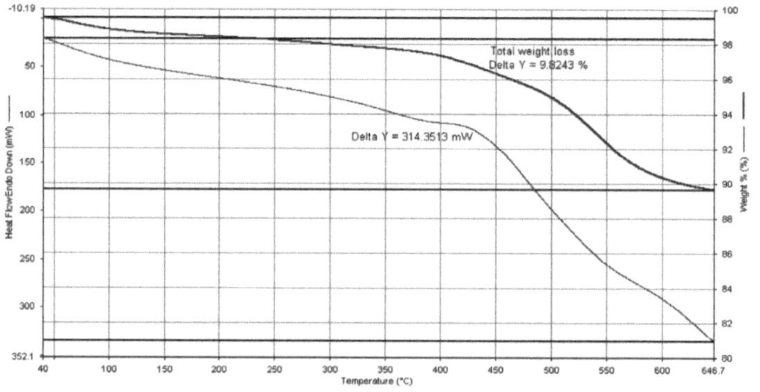

FIGURE 2.5 TGA and DTA of sample S_5.

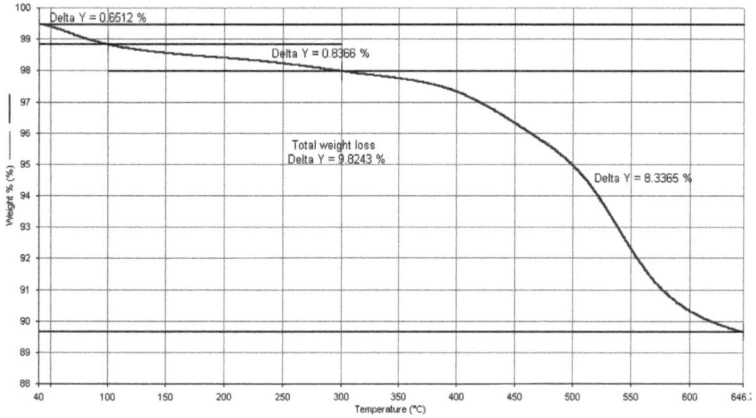

FIGURE 2.6 TGA of sample S_6.

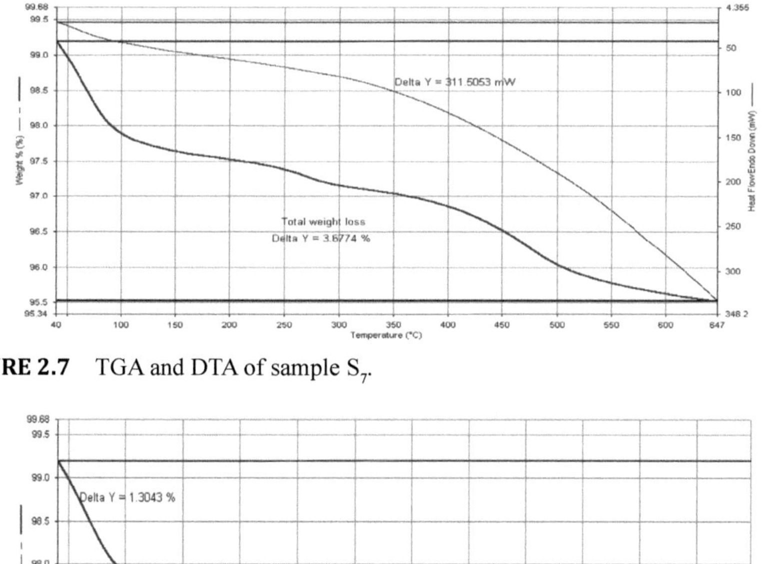

FIGURE 2.7 TGA and DTA of sample S_7.

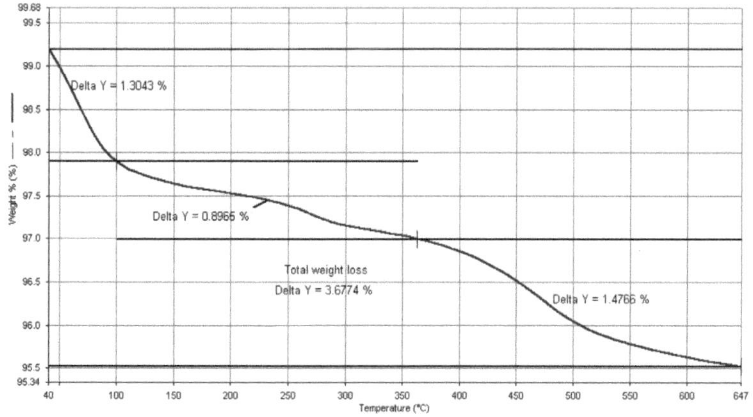

FIGURE 2.8 TGA of sample S_8.

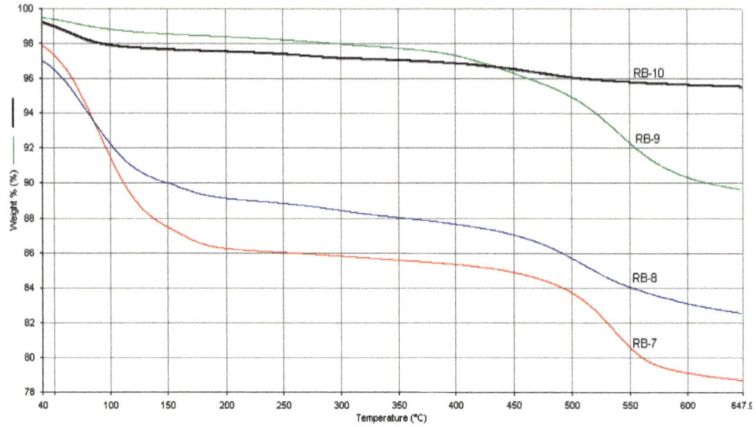

FIGURE 2.9 Comparison of TGA curves of samples, S_7, S_8, S_9, and S_{10}.

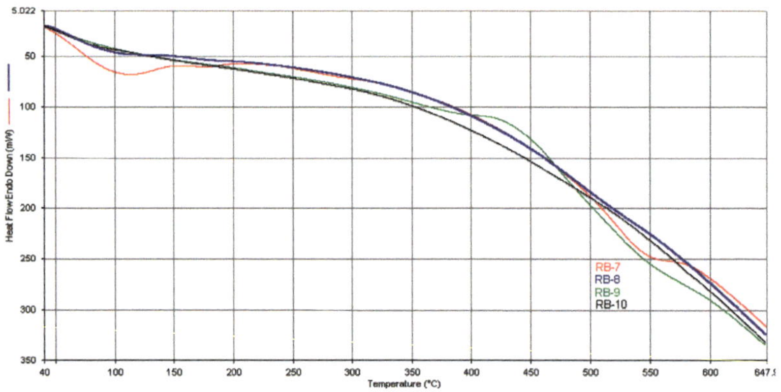

FIGURE 2.10 Comparison of DTA curves of samples, S_7, S_8, S_9, and S_{10}.

2.4 TGA AND DTA OF BENTONITE MINERAL

2.4.1 KINETICS

The kinetic study may also be done with TGA and DTA studies. In a TGA furnace, the sample is left for 2–3 min to equilibrate at the furnace temperature and then decomposition of the sample with time is followed. This process is repeated at other temperatures and the results are analyzed to determine reaction mechanisms and activation energies. However, this method has some limitations because the method is very rapid and analysis of results becomes difficult.

2.4.2 DECOMPOSITION STEPS

Multistage decomposition can be studied either by TGA or DTA. The TGA gives knowledge of individual steps by percent weight loss during decomposition of calcium oxalate monohydrate, anhydrous calcium oxalate is obtained in first step, calcium carbonate in second step, and calcium oxide in third step. A weight loss of 14.49% clearly shown in above figures shows the presence of Bentonite minerals whereas weight loss of 9.8243 shown in Figure 2.6 and weight loss of 3.674% in Figure 2.7 shows the mixed type of minerals. Thus, it can be concluded that TGA and DSC provide an important clue to know the solid nature.

2.5 SCANNING ELECTRON MICROSCOPY (SEM)

The scan electron microscope is an important tool to study the surface morphology of a material. Electron diffraction and imaging are used together in transmission electron microscopy (TEM). Images of isolated atoms can also be obtained by modern instruments. Scanning electron microscopy enables to the detection of small quantities of materials. Clear images are obtained by microscope of high resolving power. The white light of effective wavelength has the resolving power of 200 mm. Electron beams can be focused by magnetic as well as electrostatic fields.

Electrons scattered from the small irradiated area are detected and the electron's electric signal is sent to the video screen for the image.

So, microscopes of higher resolving power exist. Wavelengths of electron beams are about 10^5 times smaller than those of visible radiation. Thus, an electron microscope is capable of imaging atomic structures of the surface and adsorbates.

In SEM, a finely focused electron beam moves from one point to the next to form a pattern. This can be compared with television imaging. The intensity of scattered electrons is measured and displayed on a TV monitor to form a raster image. A surface resolution of 20–50 Å is attained with high-intensity electron sources (LaB_6). Surface SEM produces topo graphic images which can be interpreted.

One of the important features of electron microscopy is its ability to solve problems in solid-state chemistry where adsorption takes place on the surface. Palladium adsorption on charcoal can be seen in the images in a very clear way where mean particle size is around 100 Å. Adsorption of toxic heavy metals from an aqueous medium on the surface of fine powder of bentonite minerals, clay, or other clay materials can be studied with a scanning electron microscope at different magnifications.

2.5.1 SALIENT FEATURES OF SEM

i) SEM studies are used to study the structures of sizes 10^{-2} to 10^{2} μm whereas the TEM limit is around 0.1 μm.
ii) SEM is invaluable for the survey of materials under high magnification whereas TEM helps clearly to find dislocations, stacking faults, and anti-phase boundaries.
iii) A new version of SEM is FSEM, that is, field emission SEM. The FSEM has the widest range of applications and gives efficiency by enhanced surface sensitivity.
iv) SEM examines the optical and electronic properties of semiconductor materials. The SEM beam may inject charge carriers into the semiconductor. This technique is known as the beam–injection technique.
v) SEM does not provide 3D images.

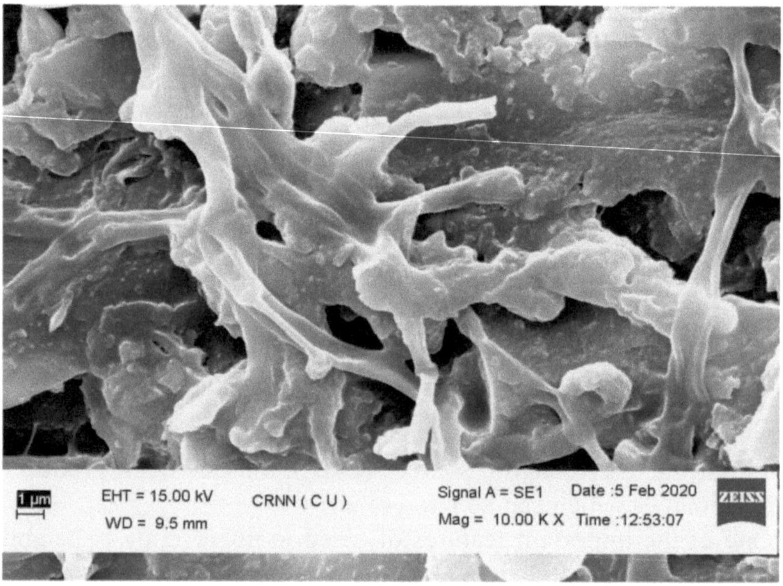

FIGURE 2.11 A SEM image of agriculture solid waste powder AKJS1_01 at a magnification of 10K×.

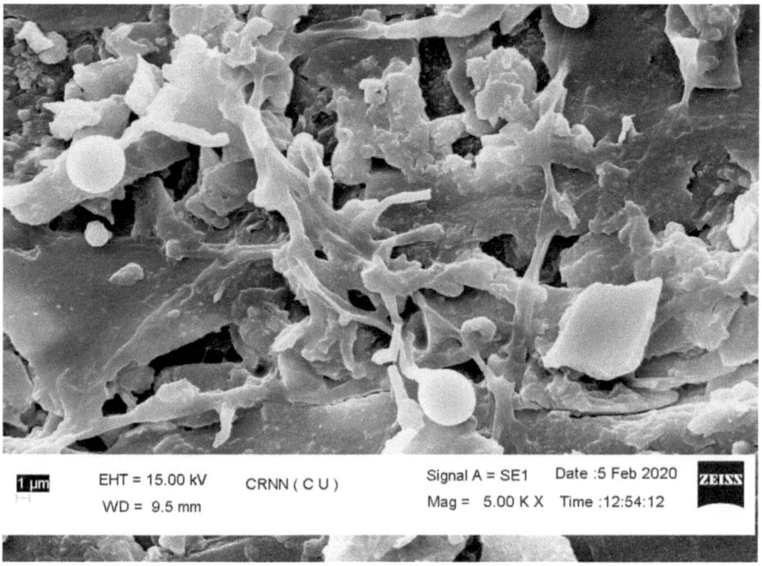

FIGURE 2.12 A SEM image of agriculture solid waste AKJS1_02 at a magnification of 5K×.

FIGURE 2.13 A SEM image of agriculture solid wasteAKJS1_03 at a magnification of 2K×.

FIGURE 2.14 A SEM image of agriculture solid waste AKJS1_04 at a magnification of 1K×.

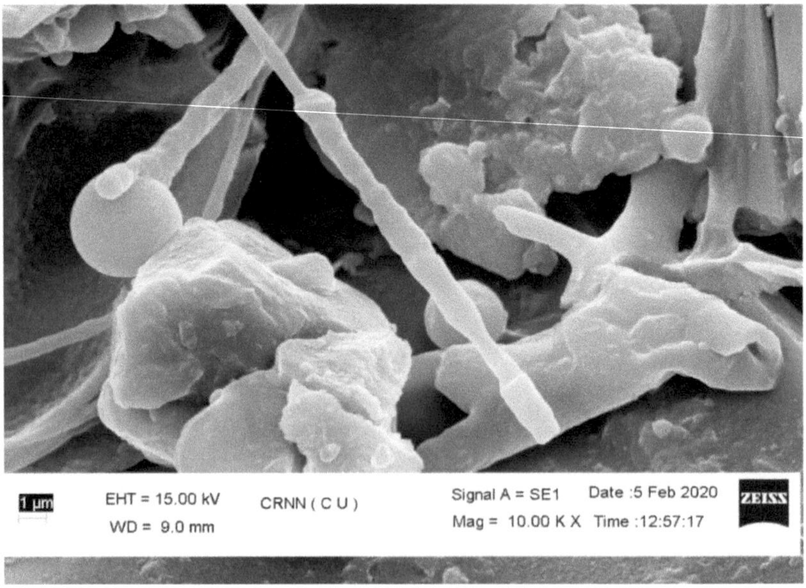

FIGURE 2.15 A SEM image of agriculture solid waste AKJS1_05.

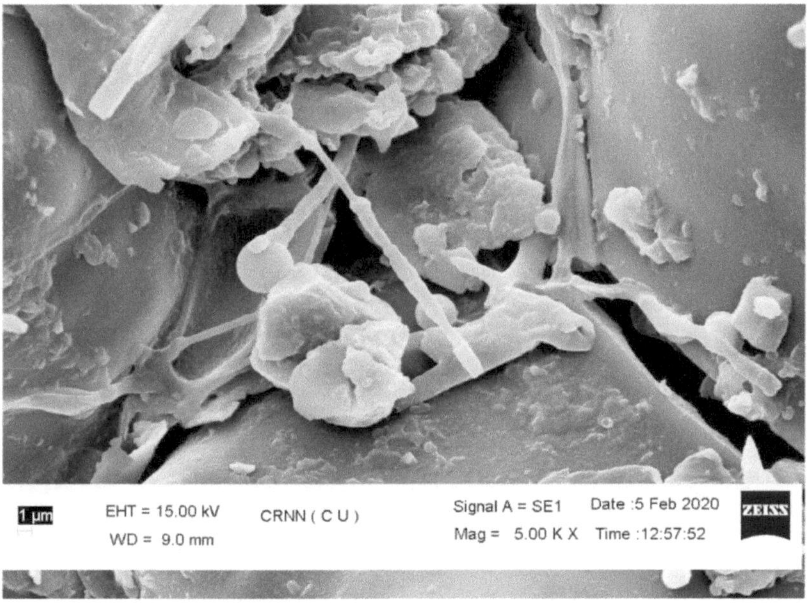

FIGURE 2.16 A SEM image of agriculture solid waste AKJS1_06.

FIGURE 2.17 A SEM image of agriculture solid waste AKJS1_09.

FIGURE 2.18 A SEM image of agriculture solid waste AKJS1_10.

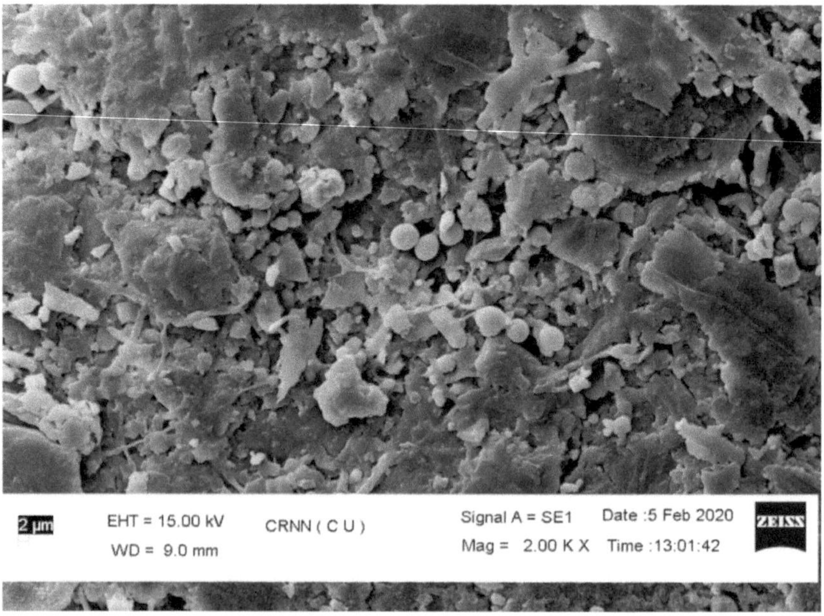

FIGURE 2.19 A SEM image of agriculture solid waste AKJS1_11.

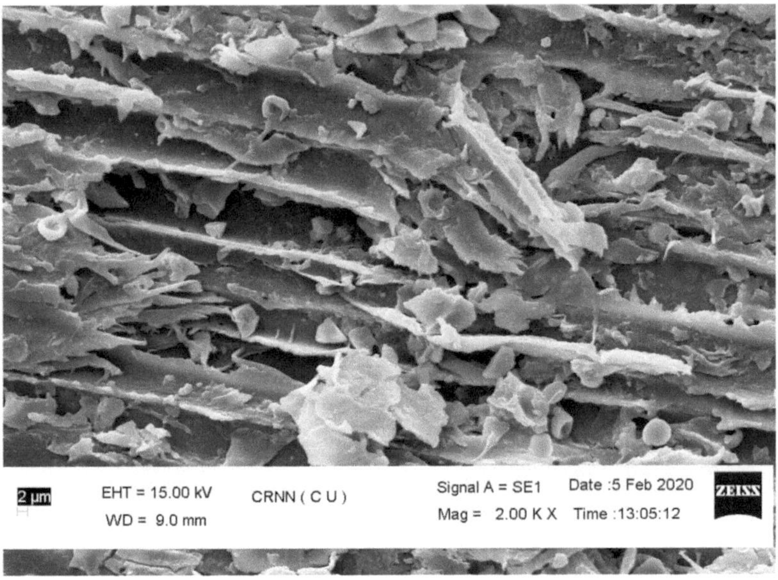

FIGURE 2.20 A SEM image of sample AKJS2_03 after adsorption of arsenic on a solid surface.

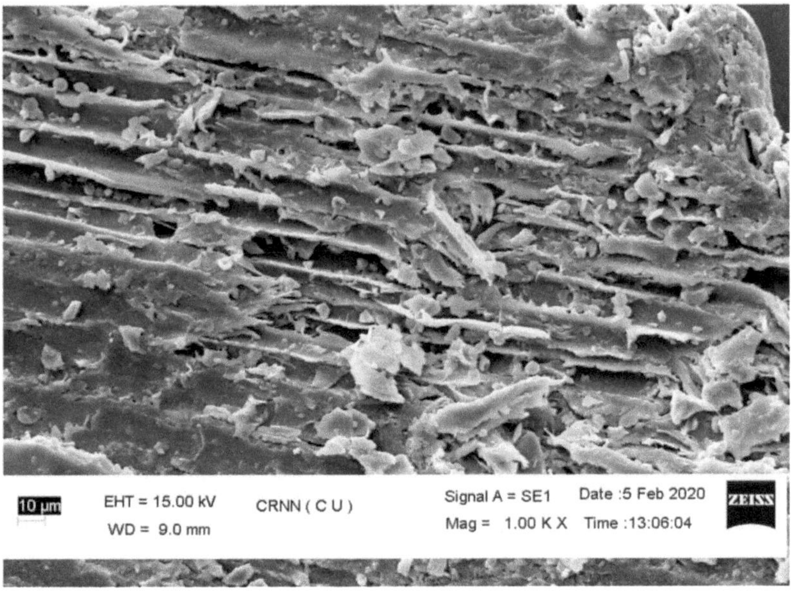

FIGURE 2.21 A SEM image of sample AKJS2_04 after adsorption of arsenic on a solid surface.

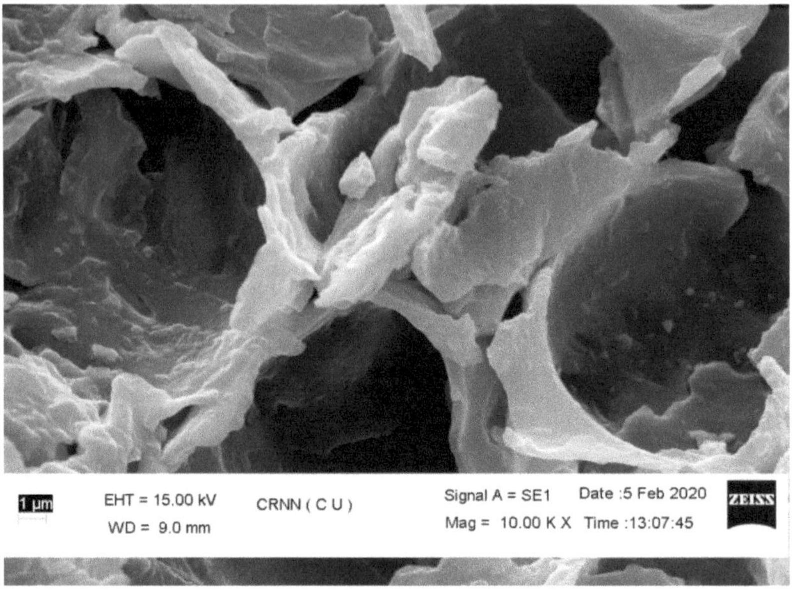

FIGURE 2.22 A SEM image of sample AKJS2_05 after adsorption of arsenic on a solid surface.

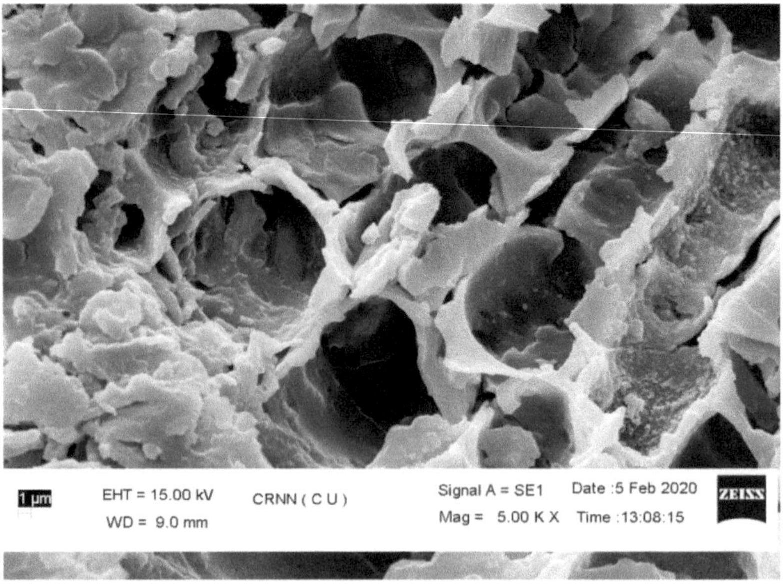

FIGURE 2.23 A SEM image of sample AKJS2_06 after adsorption of arsenic on a solid surface.

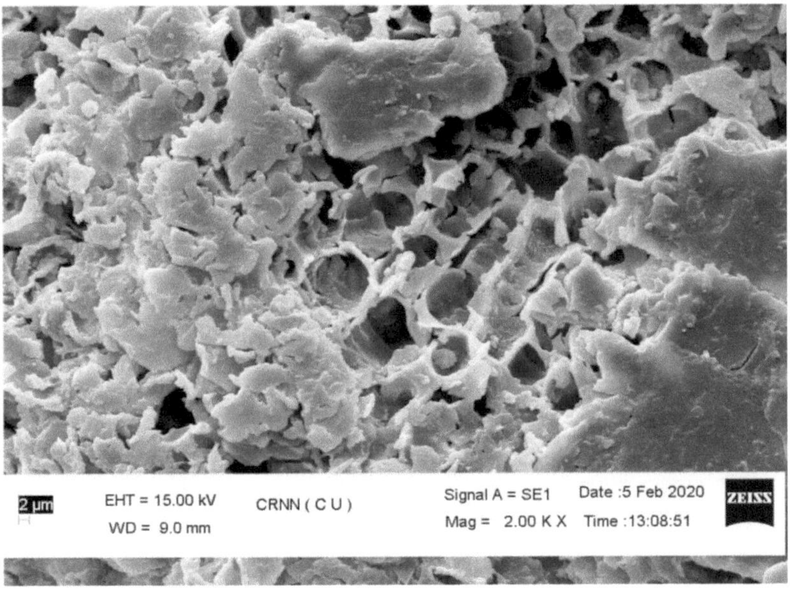

FIGURE 2.24 A SEM image of sample AKJS2_07 after adsorption of arsenic on a solid surface.

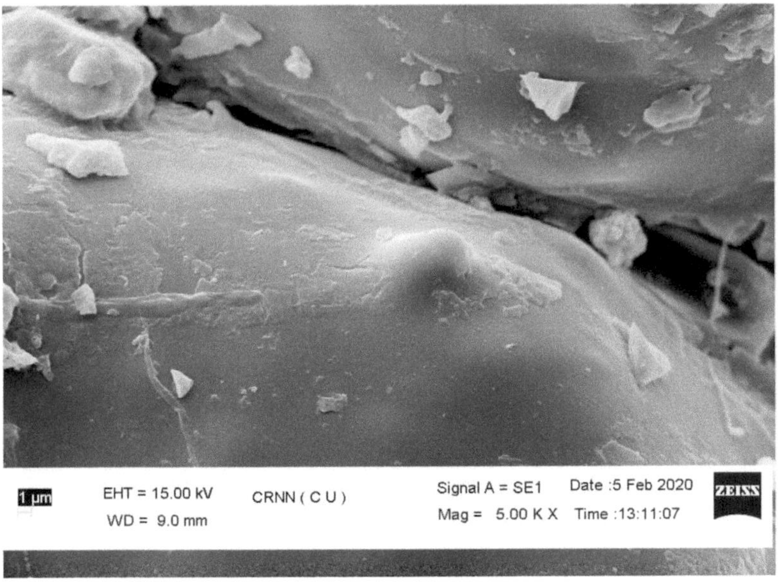

FIGURE 2.25 A SEM image of maize stem powder sample AKJS2_10.

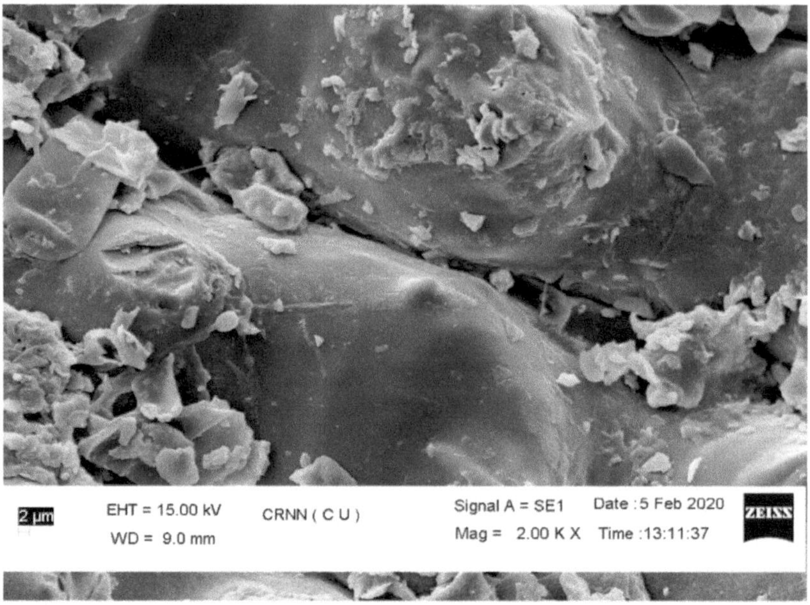

FIGURE 2.26 A SEM image of maize stem powder sample AKJS2_11.

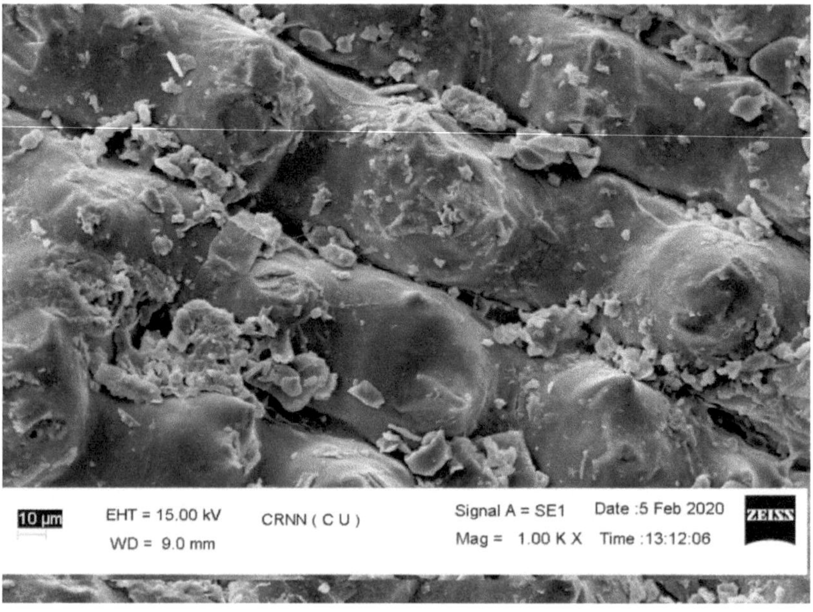

FIGURE 2.27 A SEM image of maize stem powder sample AKJS2_12.

2.6 SCANNING TUNNELING MICROSCOPY (STM)

Scanning tunneling microscopy is used for a view of the atomic structure and electron states of solids. The structure of fullerenes and layered sulfides have been studied by STM. Scanning tunneling microscopy has been invented by Binning and Rohrer et al. (1982, 1984). By STM, the topography and electronic structure of a solid surface with atomic resolution can be known. In this method, electron tunneling takes place between a sharp metal tip and the surface of a conducting solid. If the tip is brought close to the surface of the sample, the wave functions of the tip and sample overlap as a result of which electrons tunnel between the tip and surface of the solid. On application of sufficient voltage, a net flow of electrons takes place across the gap. This flow of electrons gives the tunneling current. When the bias voltage is negative, electrons tunnel from the filled electron states to the empty states in the tip. If the applied voltage is positive, electrons tunnel from the filled states in the tip to the empty states in the solid STM with. Atomic force microscopy (AFM) is used to study solid-state problems. These techniques are used to investigate metal clusters and structures of carbon nanotubes. These are also useful to study clays, silicates, and catalysts.

2.6.1 SALIENT FEATURES OF TEM

i) STM is used in the applied field and basic fundamental research with a view to getting images of solid surfaces.
ii) When solid surfaces are used as adsorbents, STM helps detect surface defects, kinks, and molecules adhering to the surface.
iii) Individual atoms are imaged and the obtained image describes the extent of adsorption on the surface.
iv) The microscopes manufactured in ultra-high vacuum are the most suitable microscopes for study of images.
v) Clean surface and corroded surface in the case of metals can be easily described by obtaining images from STM.
vi) Atomic rows sitting on top of rows of the crystal bulk are known.
vii) Even one atom thick metal grown on the terraces of crystal planes of other metal can be determined.
viii) STM is a completely new technique for the study of the structure of matter.

ix) AFM may be said a manifestation of STM. Also, AFM views the materials and surfaces more precisely.
x) STM works on quantum mechanical tunneling.
xi) STM also probes the dissociation of bonds and desorption including adsorption equilibrium.
xii) The changes in the image quality take place due to tip modification.
xiii) STM may be used to study atomic arrangements at low-index crystallographic surfaces of annealed metals.
xiv) Alloying of metals, for example, Al alloy and Fe alloy can be studied by STM and AFM. For example, fracture surface of a Fe–Si alloy can be studied by STM.

2.7 X-RAY ABSORPTION SPECTROSCOPY (EXAFS)

X-ray absorption fine structure provides information such as bond distances and coordination numbers of solid surfaces.

This is a result of K absorption of an X-ray photon and effective diffraction of the emitted electron. Photo electrons are emitted after the absorption of the X-ray photon by the absorber atoms.

Thus EDS, EDX, or XEDS is an analytical technique used for elemental analysis and the relative proportion (atomic %) of a sample. Each element has its atomic structure which allows a set of peaks on its X-ray spectrum and was recorded. The EDX elemental analysis highlights the presence of C, O, N, F, and other elements. This technique is especially helpful in studying the adsorption behavior of solid surfaces. Also, the EDX before and after adsorption can easily be compared to know the exact percentage of adsorption on the surface of the adsorbent. Fluoride, arsenic chromium, and other toxic element adsorption on the surface of clay fine powder are studied with the help of this technique.

2.8 X-RAY DIFFRACTION

X-rays are the most useful tool to determine and characterize solid. The X-ray wavelength used is the characteristic K_α radiation. The atoms or ions act as point sources to scatter the X-rays.[17–19]

Laue equation, $a\sin\phi = n\lambda$, explains the diffraction, where a is the separation of the atoms in the row, ϕ diffraction angle, and λ is the wavelength. The alternate theory of diffraction is based on Bragg's law which is adopted universally in solid-state studies.

It is known that diffraction occurs as a result of interaction between X-rays and atoms. Bragg's equation can be shown as $n\lambda = 2d \sin\theta$, where, d is the distance between two consecutive planes, θ is the Bragg's angle, and λ is the wavelength. For a given set of planes, several solutions are possible for $n = 1, 2, 3$, etc. However, we take always first-order reflection $n = 1$, d spacing for any set of planes is given by

$$\frac{1}{d^2_{hkl}} = \frac{h^2}{a^2} + \frac{k^2}{b^2} + \frac{l^2}{c^2} \tag{2.1}$$

In a cubic crystal, $a = b = c$ and for 111 plane,

$$d^2_{hkl} = \frac{a^2}{3}; d_{hkl} = \frac{a}{\sqrt{3}}$$

where θ as the glancing angle in Bragg's equation and the maximum value of $\sin\theta$ is 1, thus there is a lower limit on the observed d-spacings as

$$n\lambda = 2d \sin\theta$$

$$\lambda = 2d.1 \ (n = 1)$$

$$d = \frac{\lambda}{2}$$

$$\text{for } CuK_{\alpha}, \frac{\lambda}{2} \approx 0.77 \text{ Å}$$

If the unit cell dimensions are known, all possible d-spacings can be calculated by substituting the values of h, k, l into the formula (Eq. 2.1) mentioned above. Thus d spacings for various planes can be calculated which gives a clue to know the structure of the crystal.

2.8.1 X-RAY DIFFRACTION APPARATUS

Diffractometer X-ray diffraction experiment requires X-ray source, sample and a detector to identify the diffracted X-rays.

Monochromatic radiation wavelength λ is used and commonly used diffractometers are Debye–Scherrer Guinier (focusing), Welssenbey precession (Buerger), and automatic diffractometer. In the Weissenbery method, the film is moved parallel to the axis of rotation as the crystal rotates A screen with a narrow aperture between the crystal and the film allows reflections only from a single layer line which is recorded on the moving film. Thus, overlapping of reflections is restricted and clear crystal structure data is obtained.

2.8.2 IMPORTANCE OF INTENSITIES

Intensities of X-ray reflection play an important role in crystal structure determination. Quantitative measurements of intensity thus give a clue to test crystal structure. For this, powders should be ground to size of 1–10 μm and sieve samples prior to X-ray diffraction. If the crystal is large, it can cause error in intensity measurement. Intensities are normally expressed as peak heights in counts per minute.

Summary of structural characterization of solids:

S. No.	Structural Details	Experimental Techniques
1	Morphology, phase identification crystal plane	X-ray diffraction, electron diffraction, and electron microscopy
2	Surface morphology	STM and AFM
3	Phase transition	DTAAFM
4	Functional groups	Fourier transform infrared (FTIR) spectroscopy
5	Crystal defects	TEM

2.9 FOURIER TRANSFORMS INFRARED SPECTROSCOPY

The infrared (IR) region (4000–650 cm^{-1}) is important in studying organic compounds as well as inorganic compounds. Infrared spectrum may be regarded as falling under two regions.

i) The functional group region 4000–1300 cm^{-1}
ii) The finger print region 1300–909 cm^{-1}

The characteristic stretching frequencies for important functional groups occur in this portion of the spectrum.

Infrared absorption data for the minerals involve some impurities. The specimens exhibit bands at 3700, 3625, 3400, 1100, 1025, 910, 690, 467, and 430 cm^{-1}. There patterns are very close to halloysite. IR data along with X-ray diffraction clearly demonstrate and characterize the clay minerals such as montmorillonite, kaolinite, and illite. Bonds silicon–oxygen, MgO, Al_2O_3, etc. can easily be seen by the peaks in FTIR.

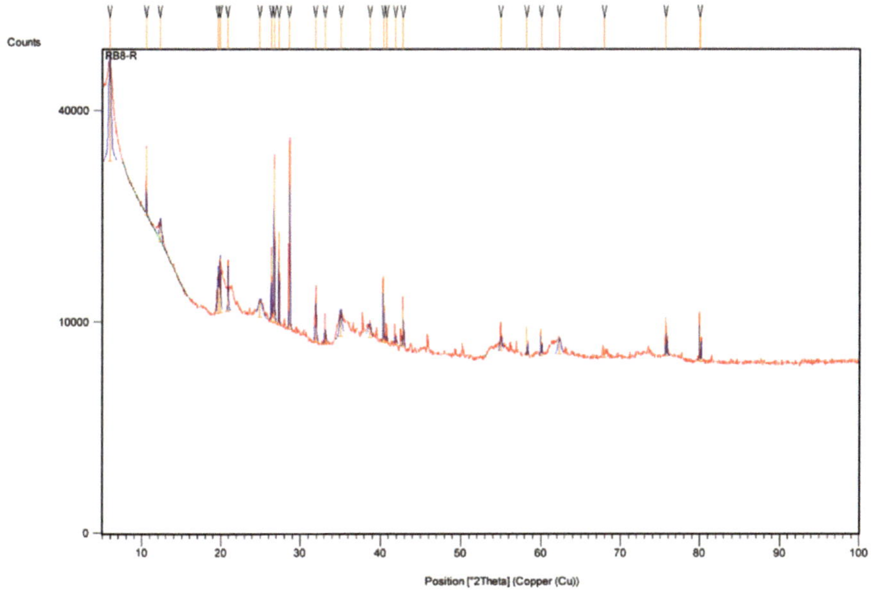

FIGURE 2.28 A PXRD of bentonite Sample 1.

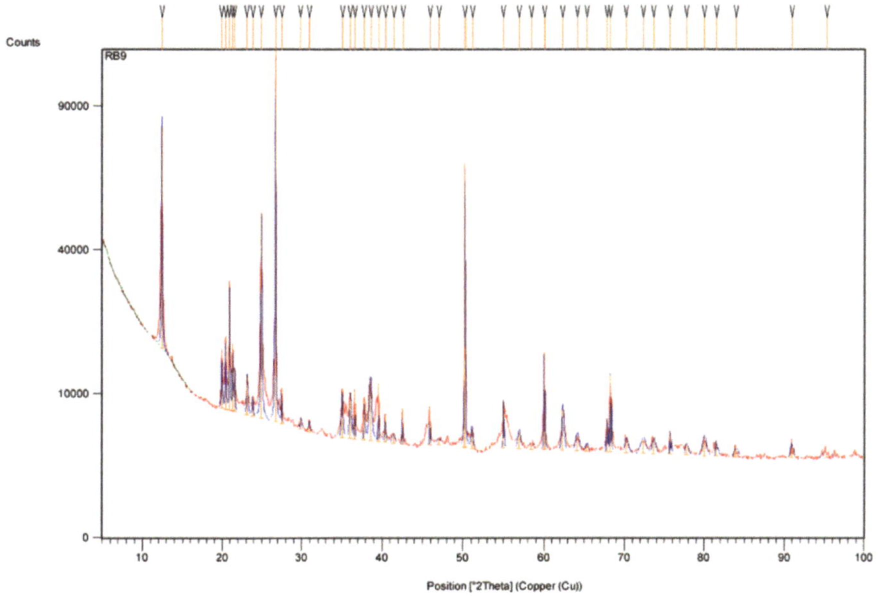

FIGURE 2.29 A PXRD of bentonite sample 2.

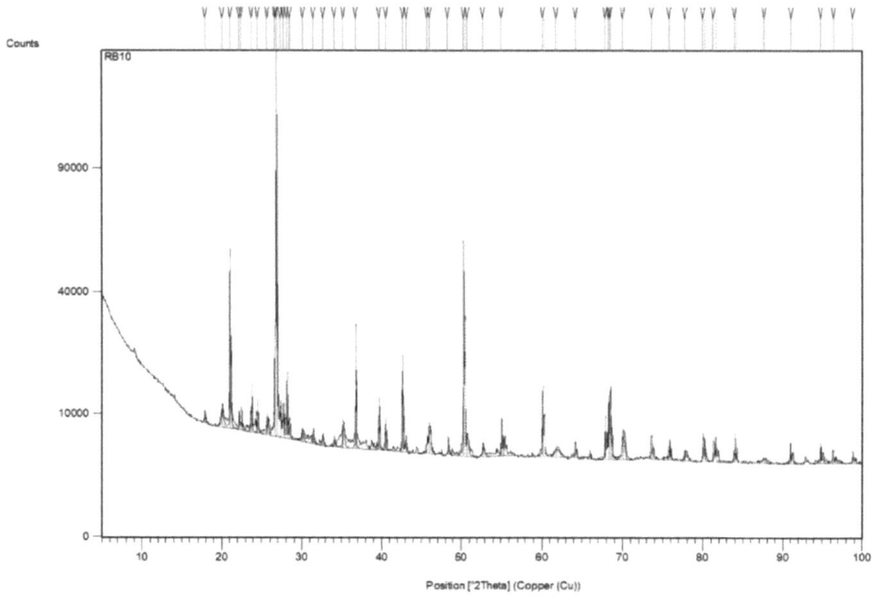

FIGURE 2.30 A PXRD of bentonite sample 3.

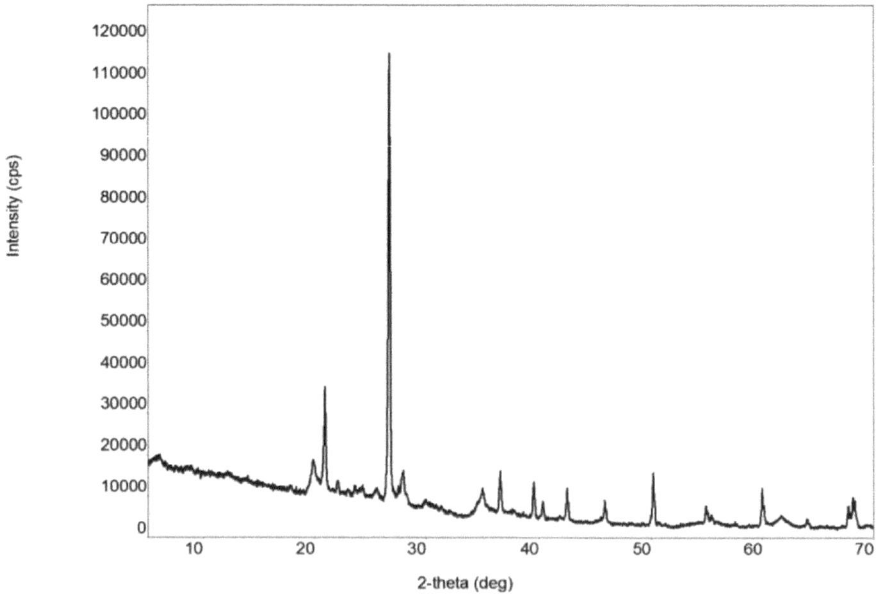

FIGURE 2.31 A PXRD of bentonite sample RHS3.

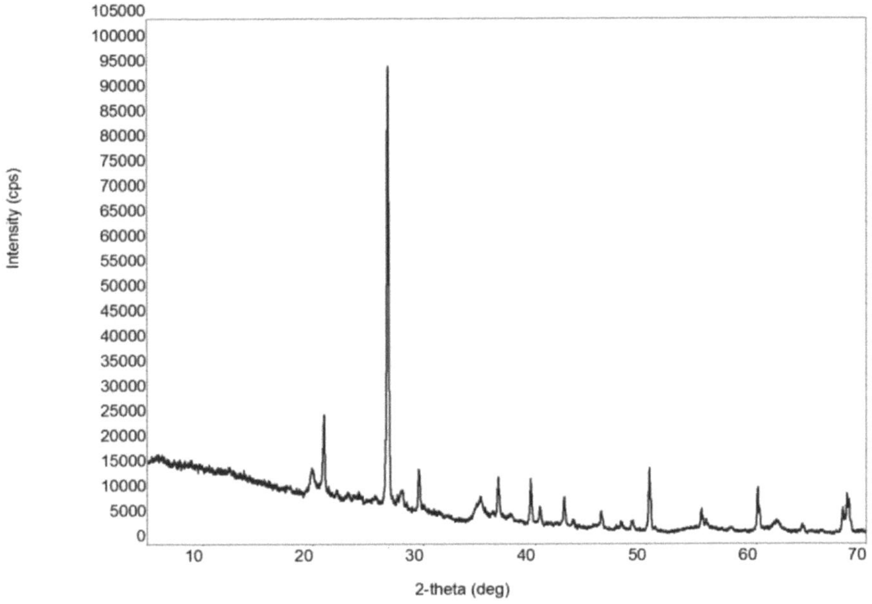

FIGURE 2.32 A PXRD of bentonite sample RHS4.

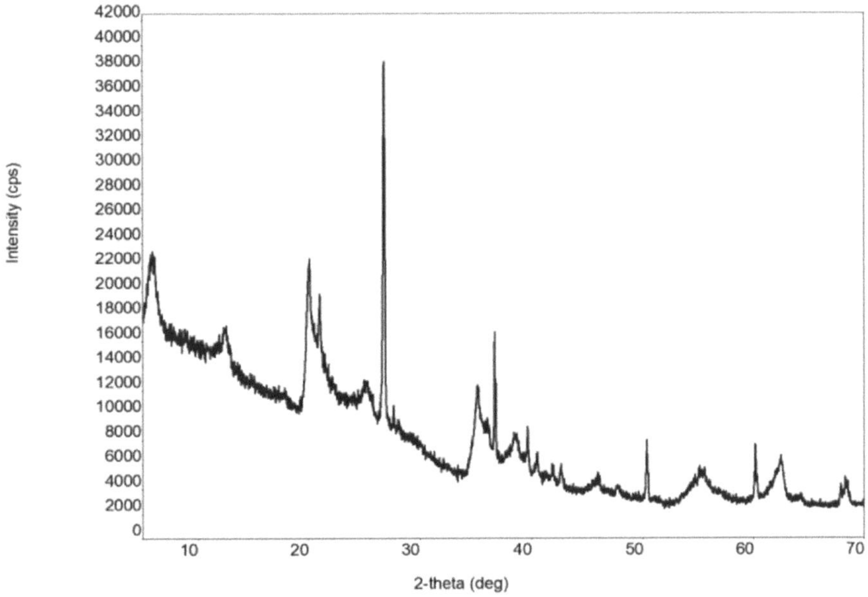

FIGURE 2.33 A PXRD of bentonite sample S_5.

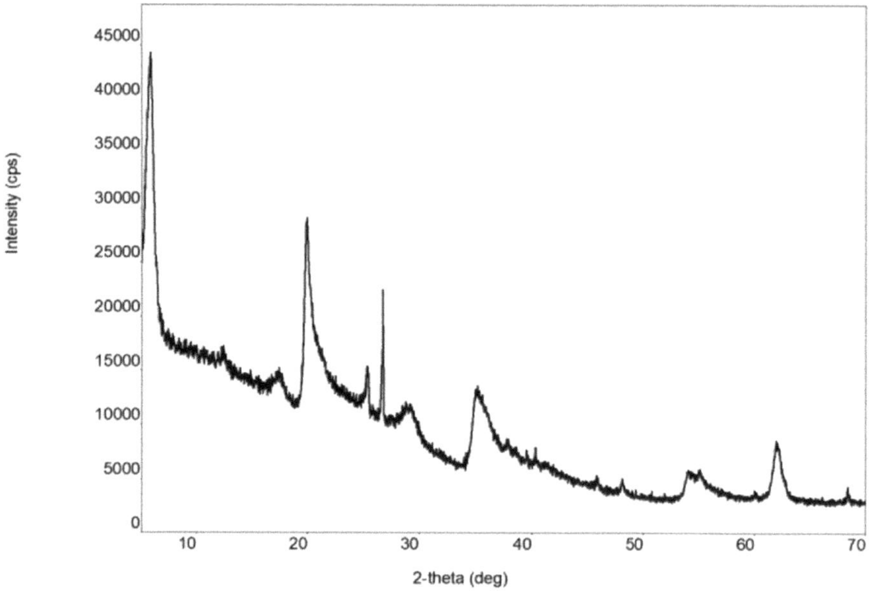

FIGURE 2.34 A PXRD of sample S_{20}.

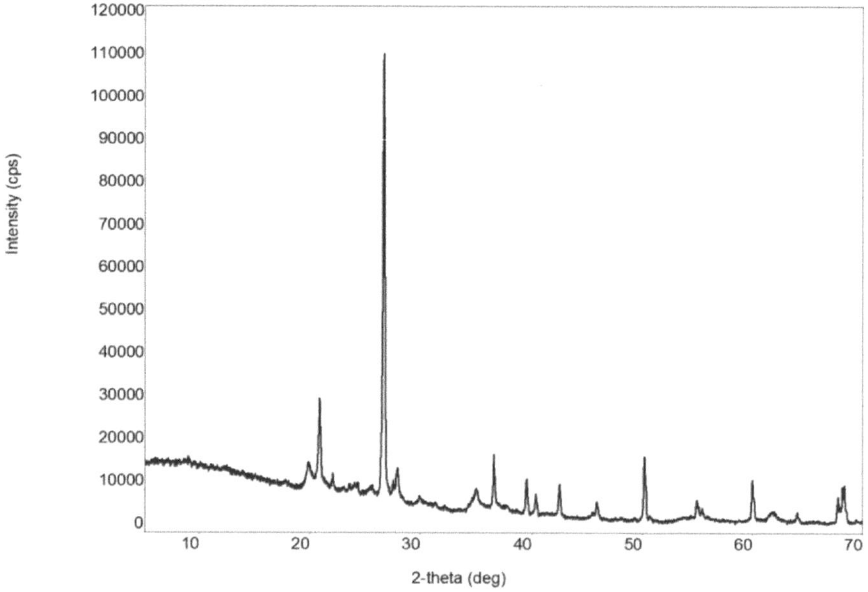

FIGURE 2.35 A PXRD of bentonite sample SH2.

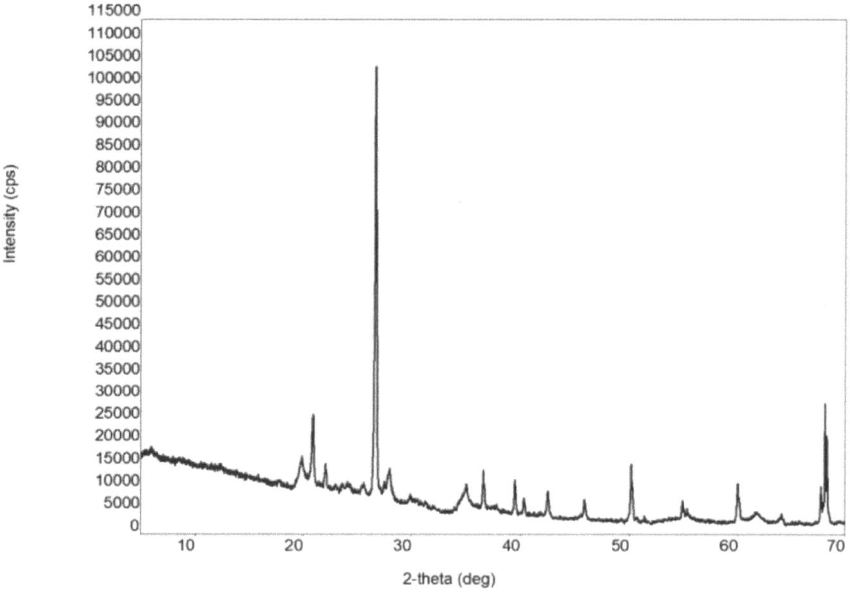

FIGURE 2.36 A PXRD of bentonite sample SH4.

The FTIR studies clearly demonstrate the adsorption of elements on the surface of the adsorbent. Adsorption of adsorbate on the surface of the adsorbent results in the disappearance of some bands, shifts and decrease in the transmission percentage. Thus, the FTIR analysis of the sorbent surface before and after the sorption process indicates the surface groups where the sorption has taken place. For example, the shift of stretching frequency from 3650 to 3451 cm^{-1} is the involvement of hydroxyl groups in the fluoride adsorption.

In a sample of Fe-rich kaolinite a first group of four bands 3692, 3669, 3654, and 3620 cm^{-1} which corresponds to the OH-stretching region and is characteristic of kaolinite.

A small band occurs at 3430 cm^{-1} which is also observed in kaolinite. In the 1200–400 cm^{-1} region, all of the bands observed are due to kaolinite. Band at 1105 cm^{-1} corresponds to silica. The band at 917 cm^{-1} corresponds to δ Al_2OH vibrations and the weak band at 881 cm^{-1} is attributed to δ FeAlOH.

The IR spectrum of rice husk powder, orange peel powder, and maize stem powder before and after adsorption has been given in the following:

Shift in bands after adsorption can be clearly seen which confirm adsorption on the solid surface.

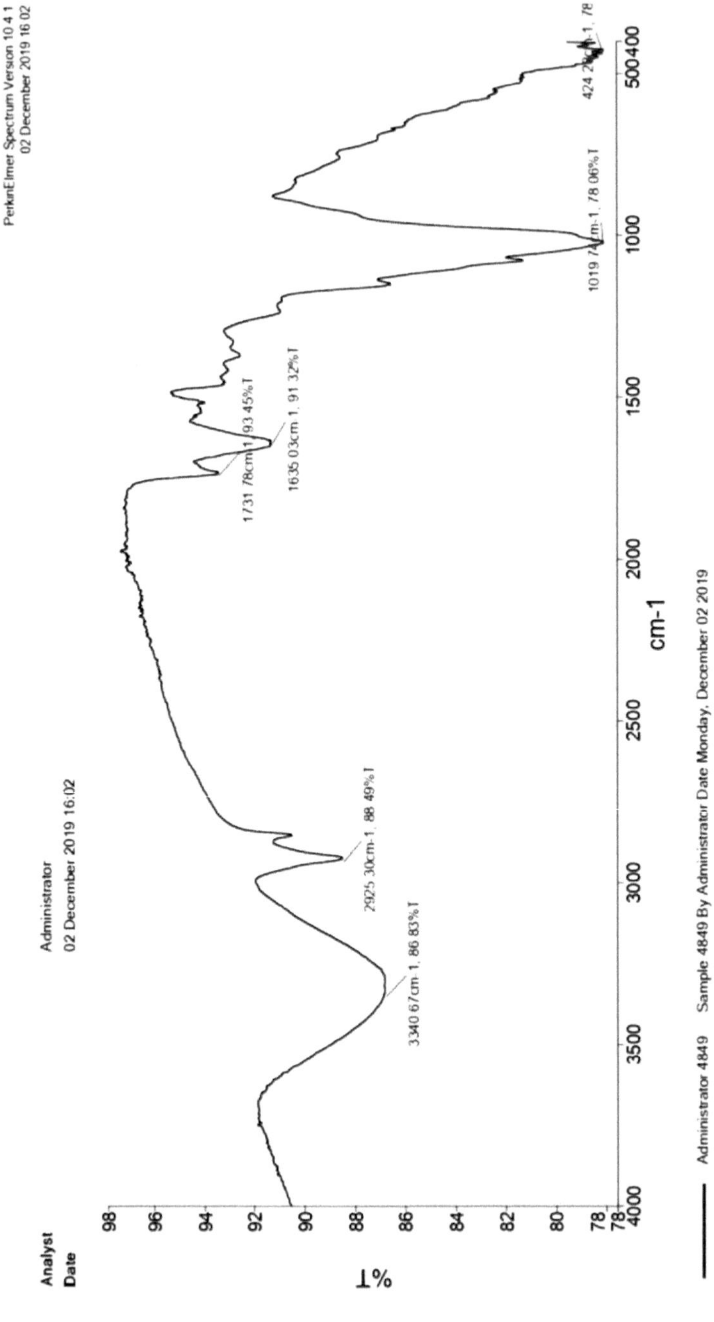

FIGURE 2.37 An FTIR spectrograph of sample S01.

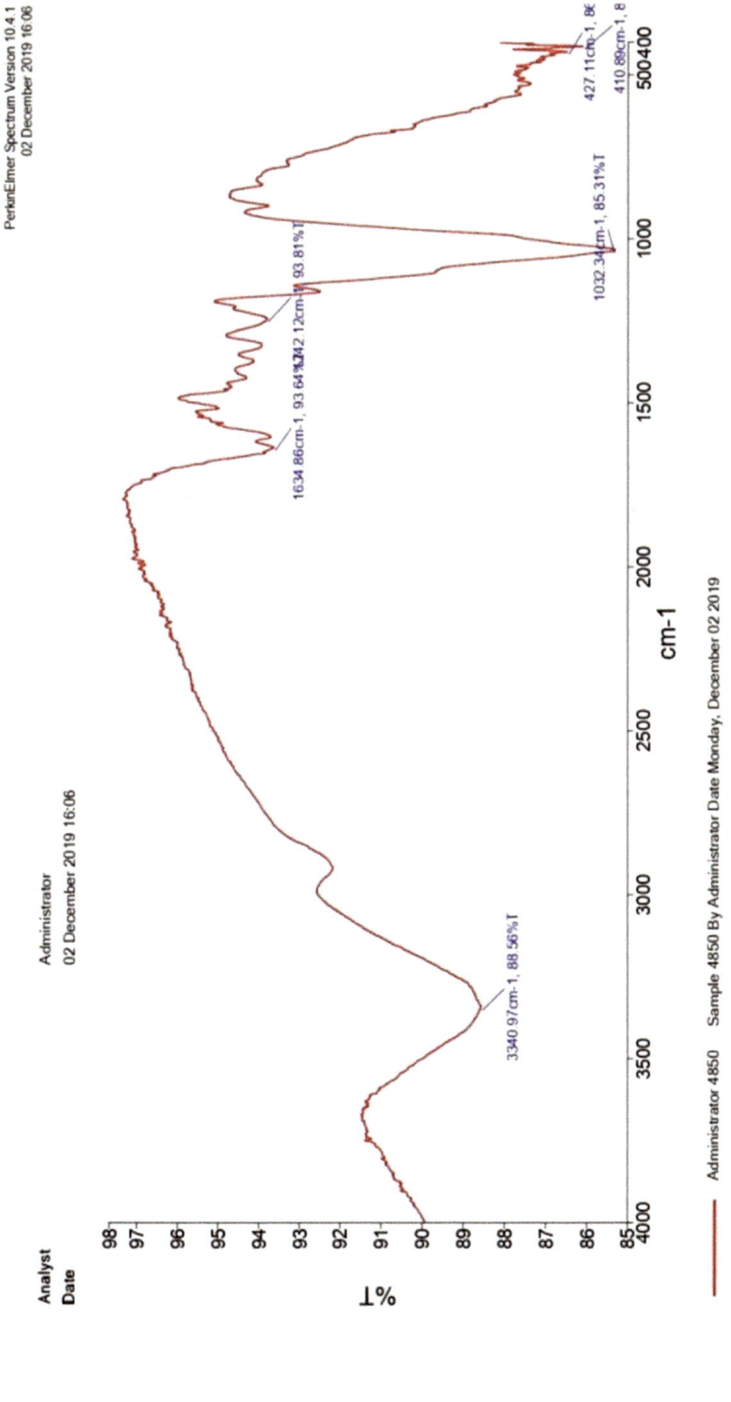

FIGURE 2.38 An FTIR spectrograph of sample S02.

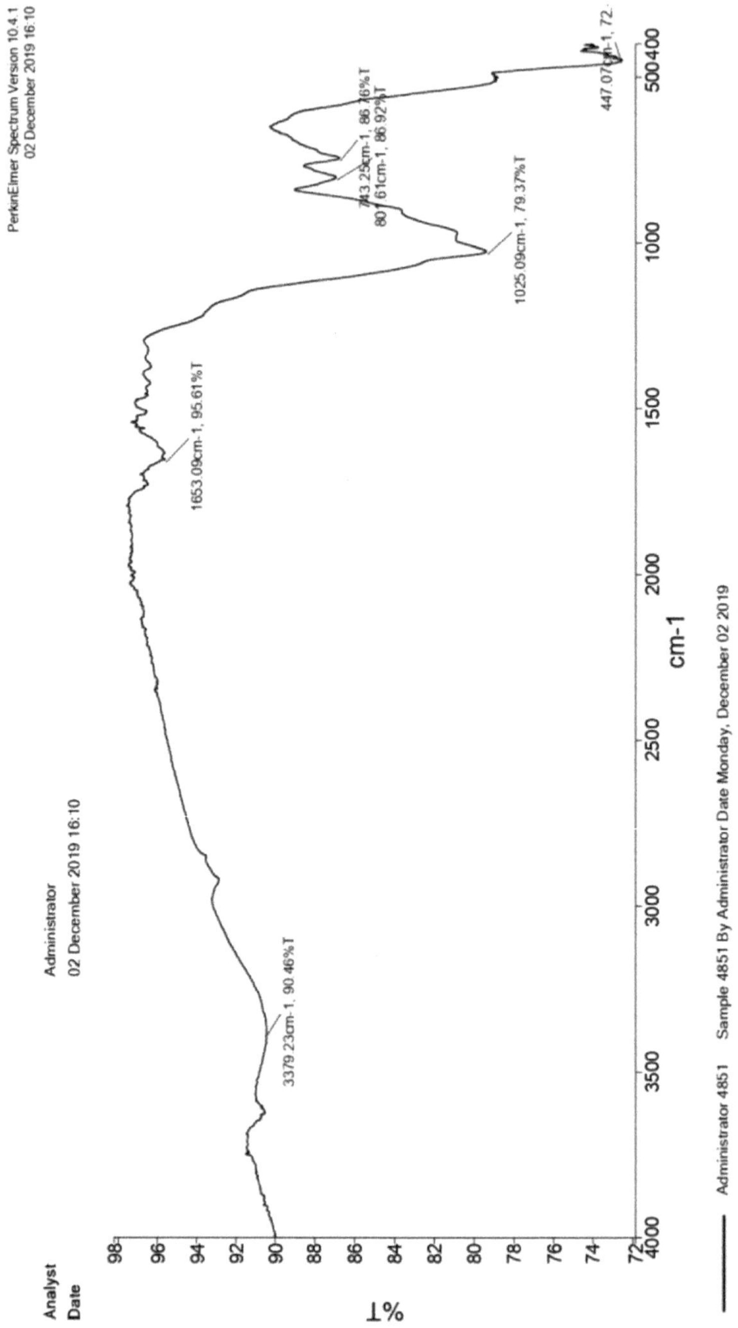

FIGURE 2.39 An FTIR spectrograph of sample S03.

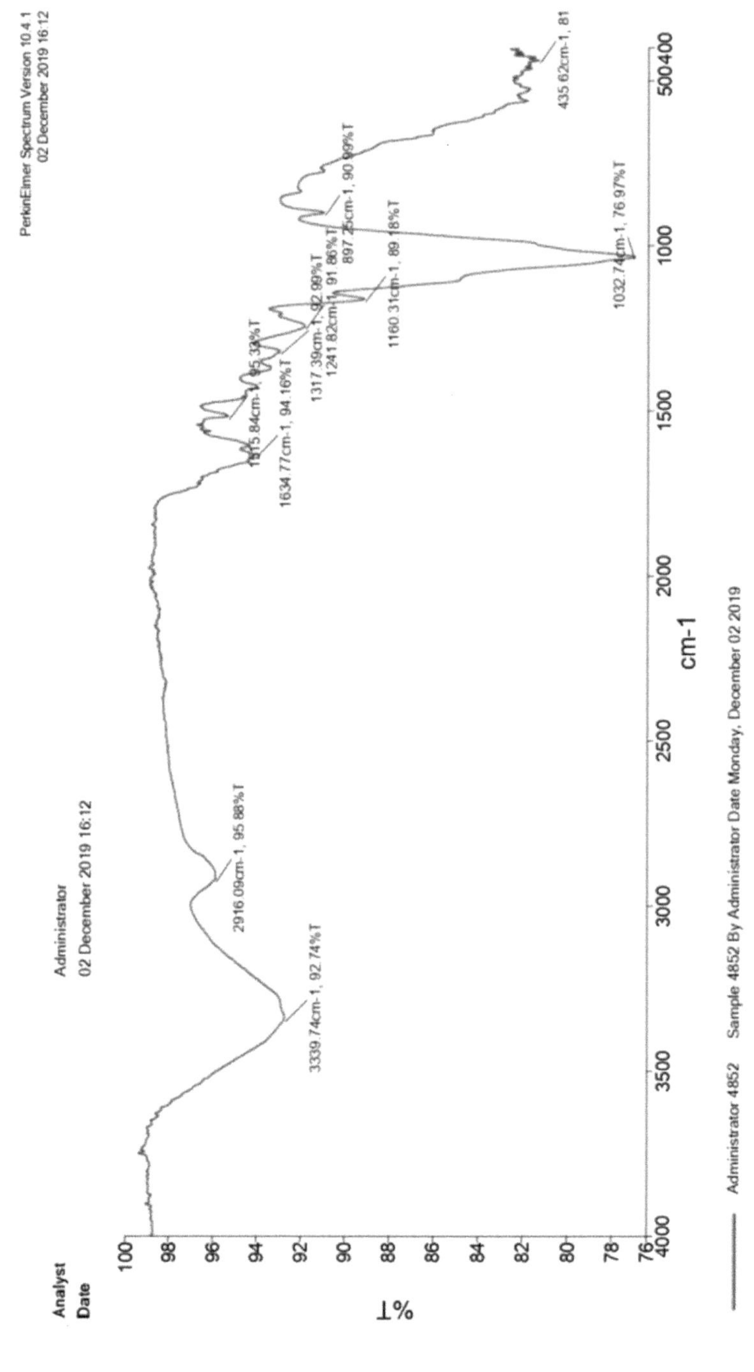

FIGURE 2.40 An FTIR spectrograph of sample S04.

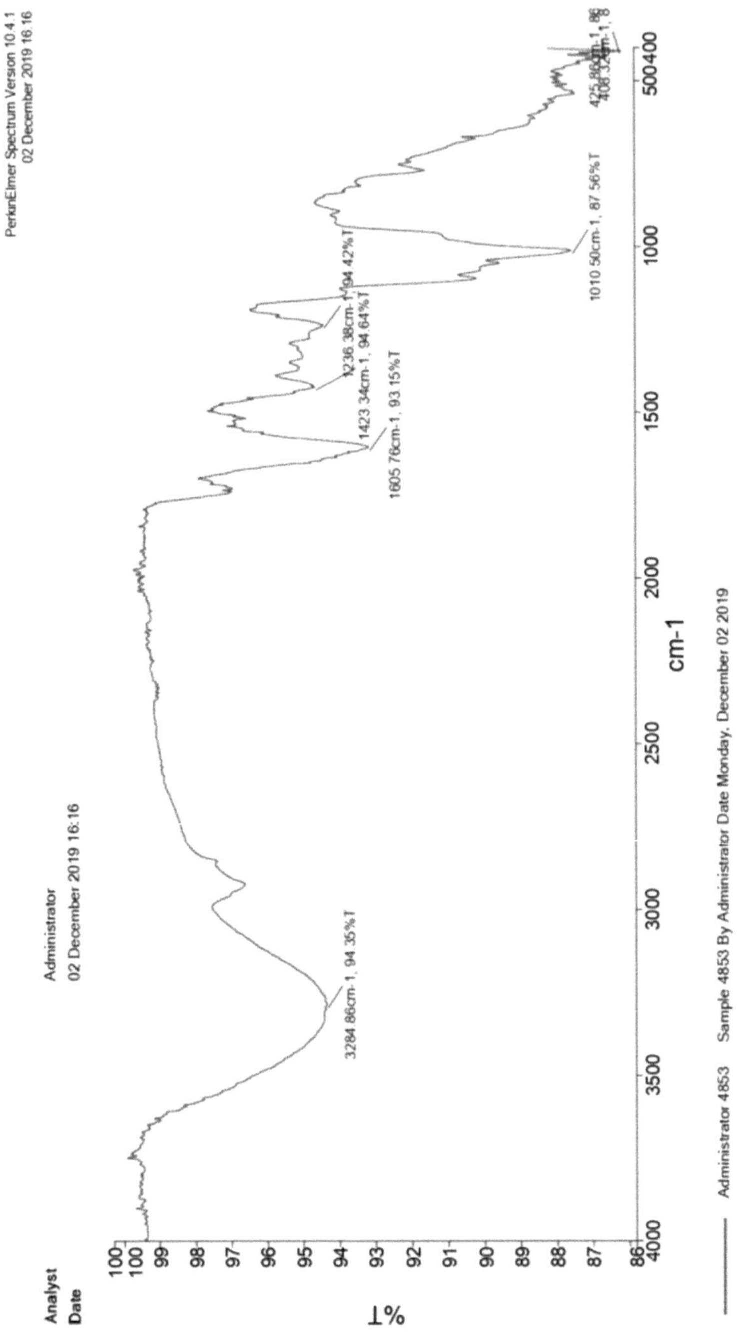

FIGURE 2.41 An FTIR spectrograph of sample S05.

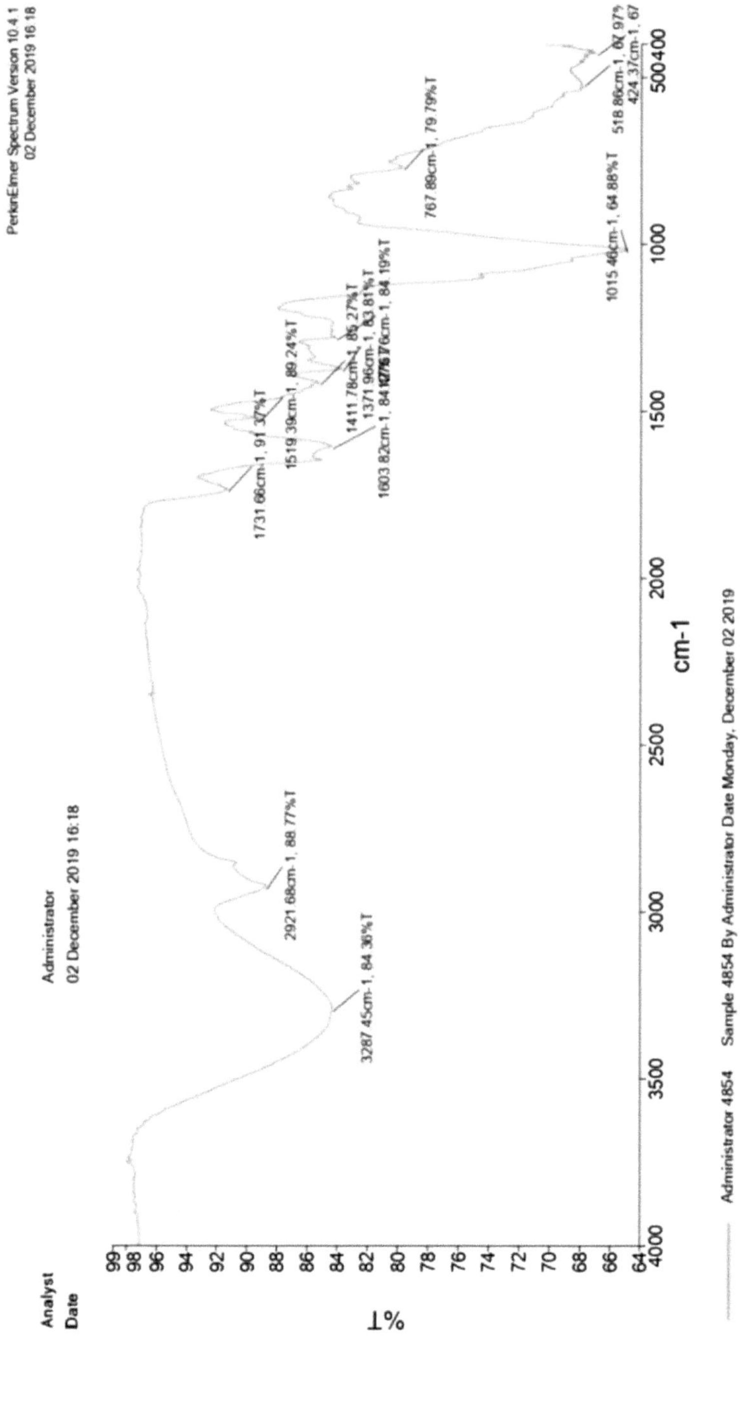

FIGURE 2.42 An FTIR spectrograph of sample S06.

Harmonic frequencies of some molecules

Molecule	ω_e(cm^{-1})
Hal	1682.6
HZn	1607.6
MgO	785.1
CaF	587.1
CaO	650
AlO	978.2
SiO	1242.0
TiO	1005
HO^-	3637
BeO	1487.3
CsCl	209.0
RbCl	228.0
MnO	840.7
FeCl	406.6
FeO	880
ZnBr	220
AgO	493.2
CuO	628.0
KCl	281.0
KBr	213.0
KI	212.0

KEYWORDS

- **applications of TGA**
- **DTA and DSC**
- **X-ray diffraction**
- **diffractometer**
- **FTIR**
- **characterization**
- **bentonite**

CHAPTER 3

Crystal Structure

ABSTRACT

Crystals are the three-dimensional arrangement of atoms. The unit cell or primitive cell is the smallest building unit of a three-dimensional lattice. The number of atoms per unit cell in a simple cubic, body centered cubic and face centered cubic are 1, 2, and 4 respectively. A lattice plane is defined with miller indices and a particular lattice plane is denoted by hkl. Powder diffractometer is used to know diffraction pattern. The structures similar to cubic ZnS are CuF, SiC, CuCl, and AgI. The mineral fluorite CaF_2 possess the structure in which eight tetrahedral interstitial sites are occupied by F^- ions and regular sites are occupied by Ca^{2+}. In the rutile structure of TiO_2, Ti is fixed at the corner and body center. $MgAl_2O_4$ is an example of spinel. The type of silicate glass possesses regular network of SiO_4 tetrahedron in a random arrangement. The most commonly used glasses are inorganic silicates,and borates. All glasses are amorphous solids or non-crystalline solids. Structure, property and application has also been dealt with.

3.1 INTRODUCTION

Three-dimensional periodic arrangement of atoms gives rise to a crystal. When crystallization occurs, identical building blocks add continuously to form a three-dimensional network. It has been discovered by mineralogist that the index numbers of orientations of the crystal faces are integers. The periodic array of similar particle gives rise to the law of integral indices.

Solid-State Chemistry: A Modern Approach. Ashok Kumar Jha, PhD (Auth.)

The lattice describes the structure of crystals in which atoms are attached to every lattice point. The basis is formed from the group of atoms. The repetition in space gives rise to the crystal structure.

3.2 LATTICE TRANSLATION VECTORS

The lattice may be defined having three fundamental translation vectors a_1, a_2, a_3 such that atomic arrangement remains unaltered from the point r

$$r' = r + u_1a_1 + u_2a_2 + u_3a_3 \tag{3.1}$$

where u_1, u_2, u_3 are arbitrary integers. The set of points r' defined by (i) for all u_1, u_2, u_3 defines a lattice.

Thus, it may be said that a lattice is a regular arrangement of atoms or ions in space.

Lattice + basis = crystal structure

Primitive translation vectors are used to define the crystal axes. The crystal axes a_1, a_2, a_3 form three adjacent edges of a parallelepiped. Any two lattice points are connected by a vector of the form given below:

$$T = u_1a_1 + u_2a_2 + u_3a_3$$

3.3 PRIMITIVE LATTICE CELL OR UNIT CELL

A primitive cell is a type of cell or unit cell or minimum-volume cell. The primitive cell always has the same number of atoms.

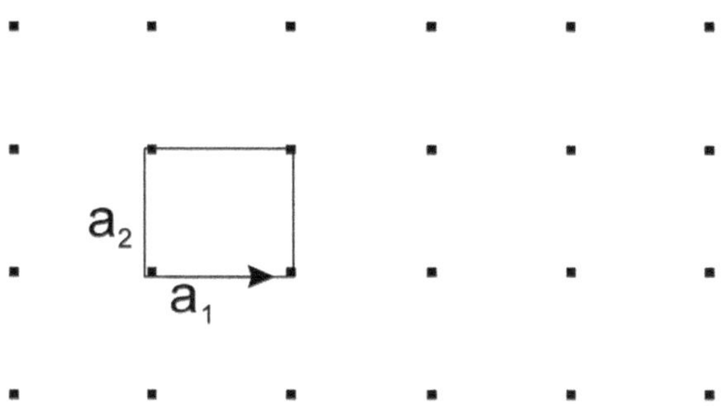

(a) Unit cell in two dimensions

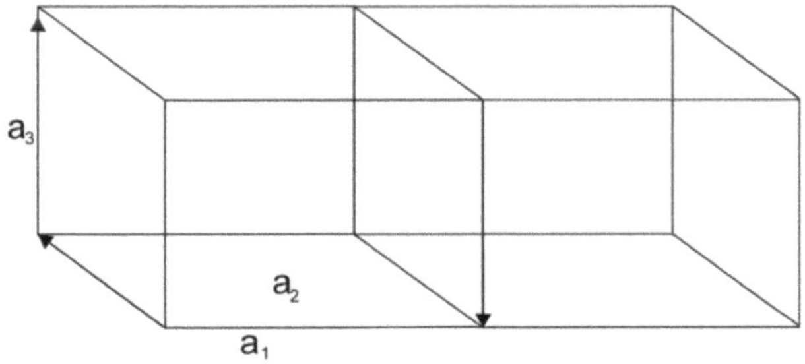

(b) Primitive cell or unit cell in three dimensions

If the unit cell is a simple cubic and lattice points are at each of eight corners, each lattice point is shared among eight cells.

The number of atoms in a unit cell in a simple cubic is $8 \times \frac{1}{8} = 1$

3.3.1 THREE-DIMENSIONAL LATTICE TYPES

Cubic, hexagonal, trigonal, tetragonal, orthorhombic, monoclinic, and triclinic are the seven basis unit cells. Table 3.1 represents lattice types in three dimensions.

TABLE 3.1 The Seven Crystal System.

S. No.	System	Axial angle	Axes	Minimum symmetry	Example
1	Cubic (SCC, BCC, FCC)	$\alpha = \beta = \gamma = 90°$	$a = b = c$	2 threefold axes	NaCl, Ag
2	Tetragonal	$\alpha = \beta = \gamma = 90°$	$a = b \neq c$	1 fourfold axis	Sn, SnO_2
3	Orthorhombic (orthorhombic (P), (orthorhombic (C), (orthorhombic (I), (orthorhombic (F),	$\alpha = \beta = \gamma = 90°$	$a \neq b \neq c$	2 twofold axis	$BaSO_4$, KNO_3
4	Monoclinic P, Monoclinic C	$\alpha = \gamma = 90°$ $\beta \neq 90°$	$a \neq b \neq c$	1 threefold axis	$NaNO_3$, ice, Sb
5	Hexagonal	$\alpha = \beta = 90°$ $\gamma = 120°$	$a = b \neq c$	1 sixfold axis	Graphite, quartz, Zn
6	Trigonal	$\alpha = \beta = \gamma < 120°$ $\neq 90°$	$a = b = c$	1 fold axis	Al_2O_3, $NaNO_3$, KNO_3
7	Triclinic	$\alpha \neq \beta \neq \gamma \neq 90°$	$a \neq b \neq c$	None	$K_2Cr_2O_7$

3.3.2 RECIPROCAL LATTICE

The reciprocal lattice is the Fourier transform of a Bravais lattice. The reciprocal lattice plays a key role in the theory of diffraction.

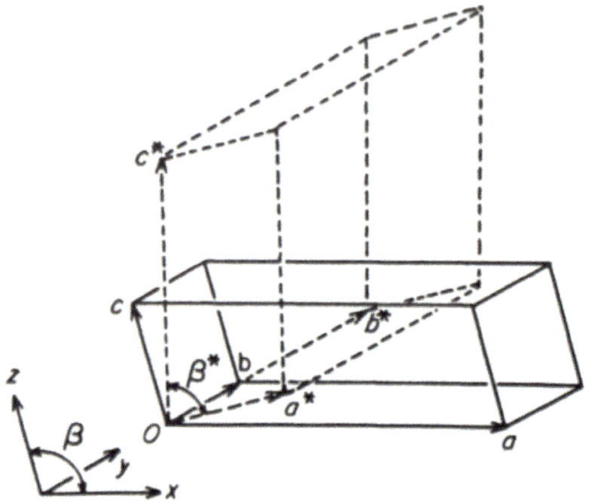

Structure of crystal system

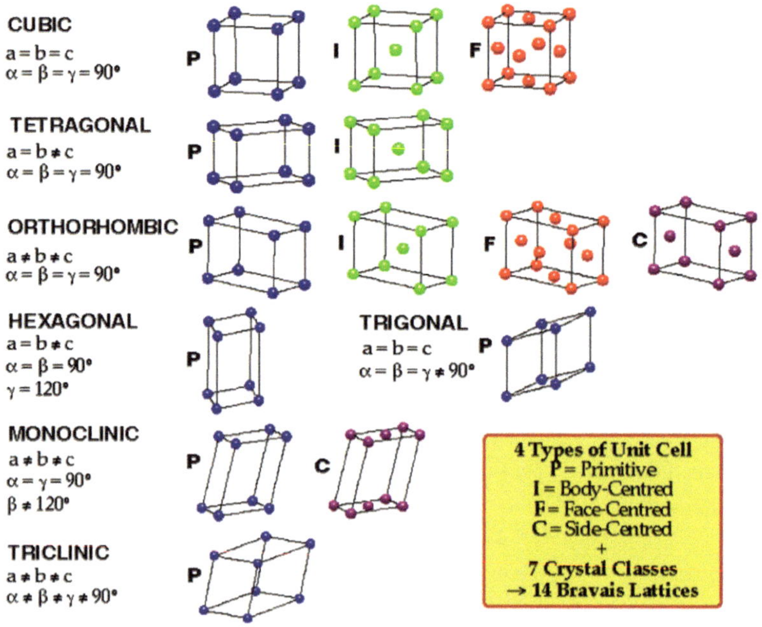

The characteristics of the three lattices in the cubic system: The simple cubic (sc) lattice, the body-centered cubic (bcc) lattice, and the face-centered cubic (fcc) lattice are given in Table 3.2.

TABLE 3.2 Characteristics of Cubic Lattices.

Volume	Simple	Body centered	Face centered
Volume, conventional cell	a^3	a^3	a^3
Lattice points per cell	1	2	4
Number of nearest neighbors	6	8	12
Nearest-neighbor distance	a	0.866 a	0.707 a
Number of second neighbors	12	6	6
Packing fraction	0.524	0.680	0.740

In the unit cell of NaCl, Nap^+ ions occupy corners and face center positions with Cl^- ions at the edge centers and body centers. The unit cell of NaCl is cubic. The three angles of the cell α, β, and γ are equal to 90° and the three edges of the cell a, b, and c are equal in length. This cubic system has certain symmetry elements.

The seven crystal systems shown in Table 3.1 are the seven independent unit cells in three-dimensional crystal structures.

3.4 THE NUMBER OF ATOMS PER UNIT CELL

In a body-centered cubic, each atom at corners of the cube contributes $\frac{1}{8}$ to the unit cell and the atom at the body contributes 1 atom to the unit cell.

So the number of atoms per unit cell = $(8 \times \frac{1}{8} + 1) = 2$ atoms per unit cell.

In a face-centered cubic, six faces of the cube are occupied by one atom each and each face atom contributes half to the unit cell. In addition to this, each atom at the corners of a cube contributes $\frac{1}{8}$ to the unit cell.

So the total number of atoms per unit cell of f.c.c is equal to $(8 \times \frac{1}{8} + 6 \times \frac{1}{2}) = 4$.

3.5 PACKING FRACTION

The ratio of the volume of sphere of a unit cell and the volume of the unit cell is known as packing fraction.

$$\text{Packing fraction} = \frac{\text{Volume occupied by atoms / ions per unit cell}}{\text{Volume of unit cell}}$$

$$\text{Packing fraction} = \frac{Z \times \frac{4}{3}\pi r^3}{a^3}$$

i) **Simple cubic lattice:** $a = 2r$, $Z = 1$

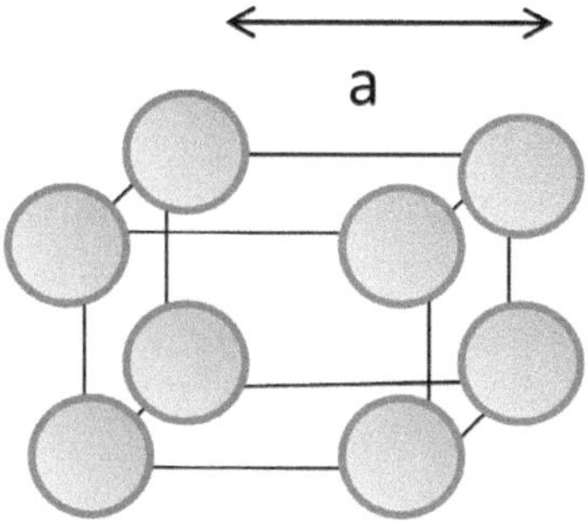

SCC Unit Cell

$$\text{Packing fraction} = \frac{1(Z) \times \frac{4}{3}\pi r^3}{a^3} = 0.52$$

Packing fraction = 52%

ii) **Body-centered cubic:** A body-corned cubic unit cell is a unit cell in which the same atoms are present at all the corners and also at the center of the unit cell.

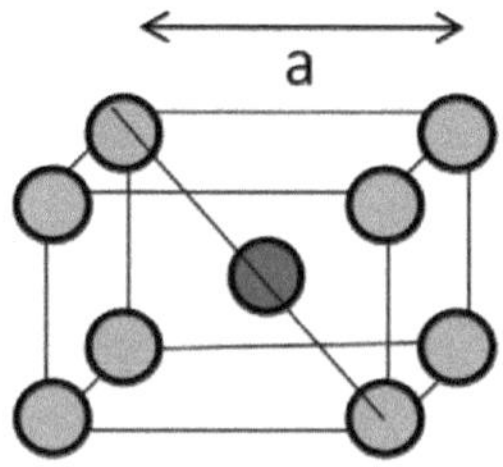

BCC Unit Cell

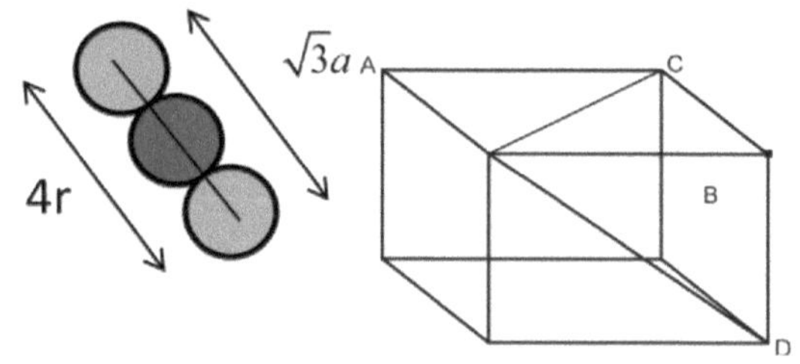

AD (Body diagonal) = $r + 2r + r = 4r$, $Z=2$

In Δ ABC

$AC^2 = AB^2 + BC^2 = a^2 + a^2 = a^2$

In Δ ADC

$AD^2 = AC^2 + CD^2 = 2a^2 + a^2 = 3a^2$

$AD = \sqrt{3a}$

$$4r = \sqrt{3}a \text{ or } a = \frac{4}{\sqrt{3}}$$

$$\text{Packing fraction} = \frac{2(Z)\times\frac{4}{3}\pi r^3}{a^3} = 0.68$$

Packing efficiency = 68% means 68% of the volume of a unit cell is occupied by the spheres.

iii) Face-centered cubic: In a face-centered cubic unit cell the atoms occupy all the eight corners of a cube and on the six faces of the cube. The number of atoms in a unit cell of f.c.c is equal to 4.

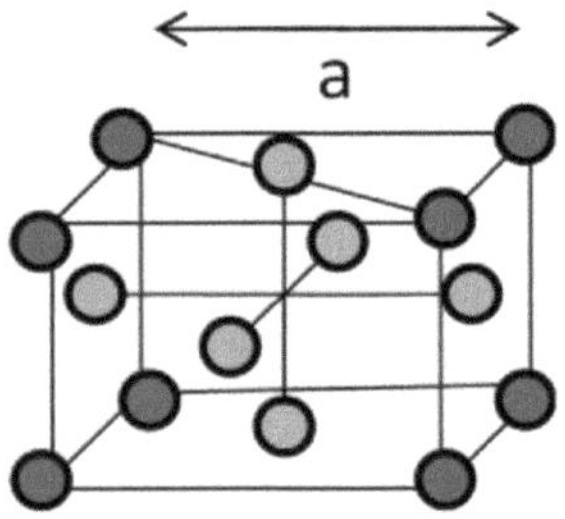

FCC Unit Cell

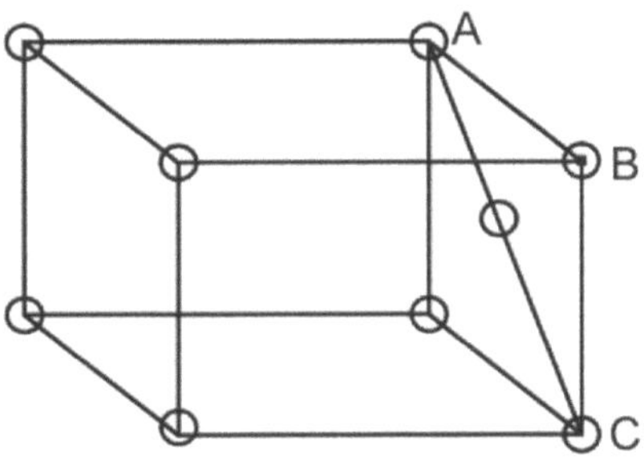

In Δ ABC,

$AC^2 = AB^2 + BC^2$

$AC^2 = a^2 + a^2$, $AC^2 = 2a^2$,

$AC = \sqrt{2a}$, $Z = 4$

$4r$ (Face diagonal) = $\sqrt{2}a$ or $a = \frac{4r}{\sqrt{2}} = 2\sqrt{2}r$

Volume of unit cell = $a^3 = \frac{4r}{\sqrt{2}} = (2\sqrt{2}r)^3$

$$\text{Packing fraction} = \frac{4(Z)\times\frac{4}{3}\pi r^3}{a^3} = 0.74$$

74% packing efficiency means 74% of the volume of a unit cell is occupied by the spheres.

iv) Hexagonal primitive unit cell:

Atom at each corner is common to six other unit cells. So their contribution to one unit cell would be $\frac{1}{6}$. The total number of atoms present per unit cell is 6 as shown below.

Three atoms in the central layer + $\frac{1}{2}$ of two atoms at face centers + $\frac{1}{6}$ of 12 atoms of the corners of the hexagon. The height of the unit cell c is $4r\sqrt{\frac{2}{3}}$ and the length of the unit cell "a" is $2r$.

The area of the base is equal to the area of six equilateral triangles

$$= 6\times\frac{\sqrt{3}}{4}a^2$$

$$= 6\times\frac{\sqrt{3}}{4}(2r)^2$$

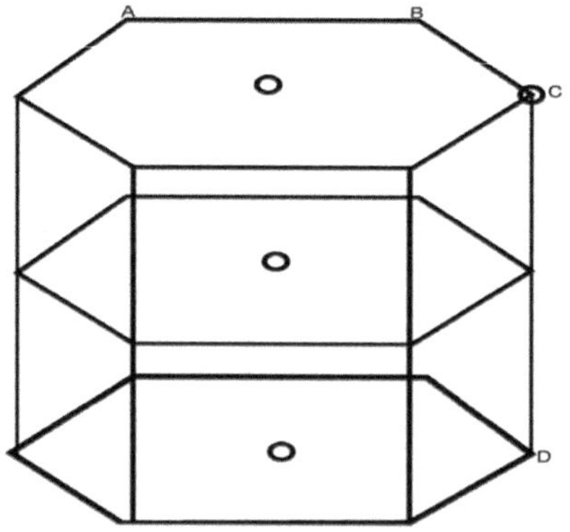

$$\text{Volume of unit cell} = 6\times\frac{\sqrt{3}}{4}(2r)^2\times 4r\sqrt{\frac{2}{3}}$$

$$\text{Volume of a sphere} = \frac{4}{3}\pi r^3,\ Z = 6$$

$$\text{Packing efficiency} = \frac{6(Z)\times\frac{4}{3}\pi r^3}{a^3} = 0.74$$

Packing efficiency = 74%

Voids = 26%

3.6 INDEX SYSTEM FOR CRYSTAL PLANES

Index system for the planes is to adopt Miller's indices. Miller's indices are the reciprocals of the coefficient of a, b, and c and the ratio is to obtain least integral values. The plane could be specified by the integral values.

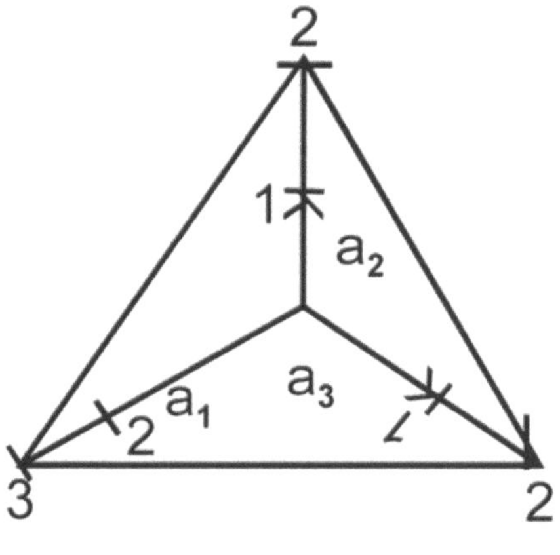

The plane intercepts are the a_1, a_2 and a_3 axes at $3a_1$, $2a_2$, $2a_3$. The reciprocals of coefficient are $\frac{1}{3}$, $\frac{1}{2}$, and $\frac{1}{3}$

The smallest ratios are now $\frac{1}{3}\times 6 : \frac{1}{2}\times 6 : \frac{1}{2}\times 6 = 2:3:3$

In general if a face of a crystal makes intercepts a_1, a_2, a_3, and the reciprocal of these numbers is hkl, the index of the plane is hkl.

If the intercepts are 4, 1, 2, the reciprocals are $\frac{1}{4}$, 1, and $\frac{1}{2}$ and smallest three integers are obtained multiplying by 4, for example, $\frac{1}{4}\times 4 : 1\times 4 : \frac{1}{2}\times 4 = 142$ plane.

For infinity, the corresponding index is zero, for example, $\frac{1}{\infty} = 0$

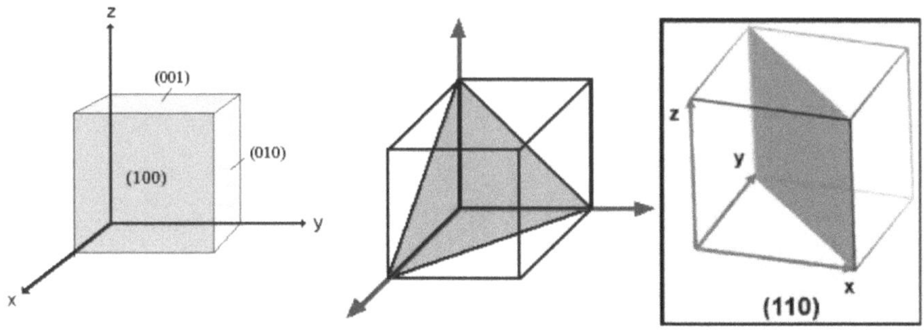

(111) plane

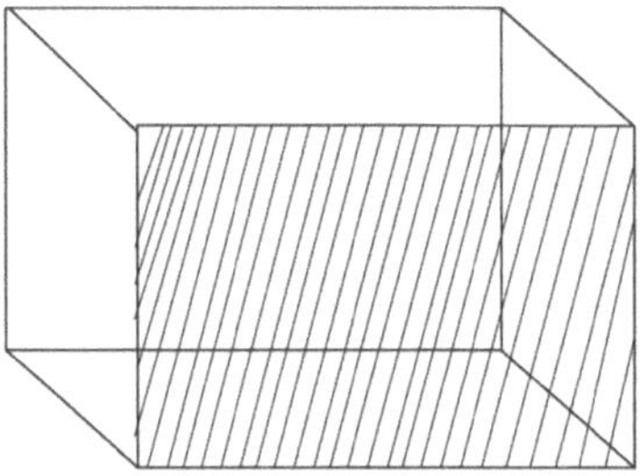

$\bar{1}00$ Plane

If the plane cuts the axis on the negative side of the origin, the corresponding index is negative expressed by putting a minus sign above the index hkl

3.7 DISTANCE BETWEEN CONSECUTIVE PLANES

A lattice plane is defined with miller indices, *h*, *k*, *l*. A particular lattice plane is denoted by (*hkl*). Other planes parallel to this plane will have the same index. The perpendicular distance d_{hkl} between two consecutive parallel lattice planes is given by

$$\frac{1}{d^2_{hkl}} = \frac{h^2}{a^2} + \frac{k^2}{b^2} + \frac{l^2}{c^2}$$

In a cubic lattice $a = b = c$

$$d_{hkl} = \frac{a}{\sqrt{h^2 + k^2 + l^2}}$$

For 100 plane in NaCl lattice

$$d_{100} = \frac{a}{\sqrt{1^2 + 0^2 + 0^2}} = \frac{a}{1} = a$$

For 110 plane in NaCl lattice

$$d_{110} = \frac{a}{\sqrt{1^2 + 1^2 + 0^2}} = \frac{a}{\sqrt{2}}$$

For 111 plane in NaCl lattice

$$d_{111} = \frac{a}{\sqrt{1^2 + 1^2 + 1^2}} = \frac{a}{\sqrt{3}}$$

Then $d_{100} : d_{110} : d_{111} = a : \frac{a}{\sqrt{2}} : \frac{a}{\sqrt{3}} = 1 : \frac{1}{\sqrt{2}} : \frac{1}{\sqrt{3}}$

It is easy to index crystals of higher sum symmetry, for example, cubic, tetragonal, or hexagonal.

It is clear that some of the (*hkl*) planes will be missing from the diffraction pattern. In principle reflection from all planes are possible for a simple cubic crystal. If ($h + k + l$) is an even number for the reflections, the crystal is a body-centered cubic. A face-centered cubic crystal gives only these reflections for which the (*hkl*) value is either all odd or even.

For tetragonal crystal, $a = b \neq c$

$$\frac{1}{d^2_{hkl}} = \frac{h^2}{a^2} + \frac{k^2}{b^2} + \frac{l^2}{c^2}$$

$$\frac{1}{d^2_{hkl}} = \frac{h^2 + k^2}{a^2} + \frac{l^2}{c^2} \quad \because a = b$$

If $l = 0$ for a plane

$$\frac{1}{d^2_{hkl}} = \frac{h^2 + k^2}{a^2} + \frac{0^2}{c^2} = \frac{h^2 + k^2}{a^2}$$

$$\text{or } d^2_{hkl} = \frac{a^2}{h^2 + k^2}, d_{hkl} = \frac{a}{\sqrt{h^2 + k^2}}$$

3.8 POWDER METHOD DIFFRACTOMETER

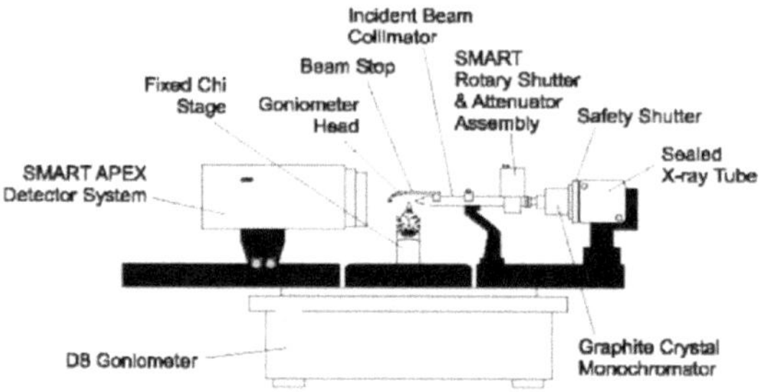

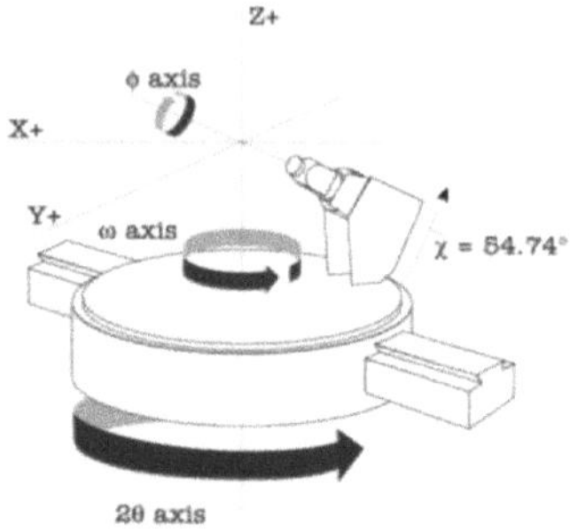

Crystal structure may be determined by locating the coordinates. This method was developed by a Debye Scherer camera that was a photographic film to detect the X-ray beam. The crystal sample is taken in a thin wetted glass capillary. A beam of homogeneous X-rays is allowed to fall on the powder. Minute crystals are oriented in all possible directions in the powdered sample. Some of the crystals are oriented in such a way that it produces maximum reflections from the 100 plane. Some produce maximum reflection from the 110 and 111 planes respectively. The diffracted beam takes the shape of a cone. Such concentric planes are formed for different lattice planes when these beam cones fall on a photographic film – strip, each cone makes the image of two areas or lines where the cone cuts the film. Thus, each diffraction cone from a set of miller planes in the crystal forms a pair of lines on the film.

If θ is the diffraction angle for a set of parallel planes, the angle between the undeflected beam and the diffracted ray will be 2θ.

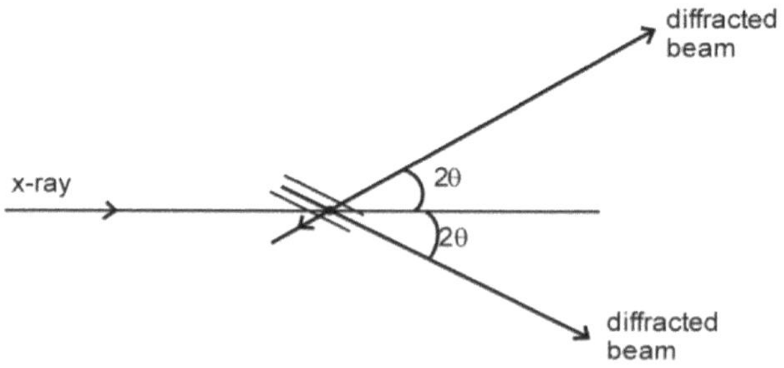

Now the diffracted beam will describe a cone with angle 4θ.

If a pair of lines are separated by a distance S and R is the radian of circled film,

$$4\theta = \frac{S}{R}\ radian$$

$$4\theta = \frac{S}{R} \times \frac{180}{\pi} \deg ree$$

If S is measured, the corresponding values of θ are obtained.

The interplanar distance d can be calculated by using Bragg's equation $n\lambda = 2d\sin\theta$

3.8.1 POWDER DIFFRACTOMETER

For this purpose powder diffractometer is used. The powdered sample is taken in the form of a thin circular layer of about 15 mm diameter and 1–2 mm thickness. The X-ray beams fall in the sample. The X-ray detector converts the intensity into current and the diffraction angle 2θ is plotted against intensity.

The d-spacings of the 100, 110, and 111 planes for all the three types of cubic lattices can be measured geometrically

$$d_{100} : d_{110} : d_{111}$$

Simple cubic Lattice: $a : \frac{a}{\sqrt{2}} : \frac{a}{\sqrt{3}} = 1 : \frac{1}{\sqrt{2}} : \frac{1}{\sqrt{3}} = 1 : 0.707 : 0.577$

Body-centered cubic: $\frac{a}{2} : \frac{a\sqrt{2}}{2} : \frac{a}{2\sqrt{3}} = \frac{1}{2} : \frac{1}{\sqrt{2}} : \frac{1}{2\sqrt{3}} = 1 : 1.414 : 0.577$

Face-centered cubic: $\frac{a}{2} : \frac{a}{2\sqrt{2}} : \frac{a}{\sqrt{3}} = \frac{1}{2} : \frac{1}{2\sqrt{2}} : \frac{1}{\sqrt{3}} = 1 : 0.707 : 1.154$

Experimental determination of the inter planar distance ratio is examined with respect to the geometrically known interplanar distance ratio.

If we consider first-order reflection according to the Braggs equation, we have

$$d_{100} : d_{110} : d_{111} = \frac{\lambda}{\sin 5.9^o} = \frac{\lambda}{\sin 8.4^o} = \frac{\lambda}{\sin 5.2^o},$$

$$d_{100} : d_{110} : d_{111} = \frac{1}{0.103} : \frac{1}{0.146} : \frac{1}{0.091} = 1 : 0.705 : 1.14$$

This ratio is the same as geometrically derived for face-centered cubic. So NaCl is a face-centered cubic. The angles for first-order reflection for the 100, 110, and 111 planes are 5.9°, 8.4°, and 5.2°, respectively.

3.9 BRAGG'S LAW

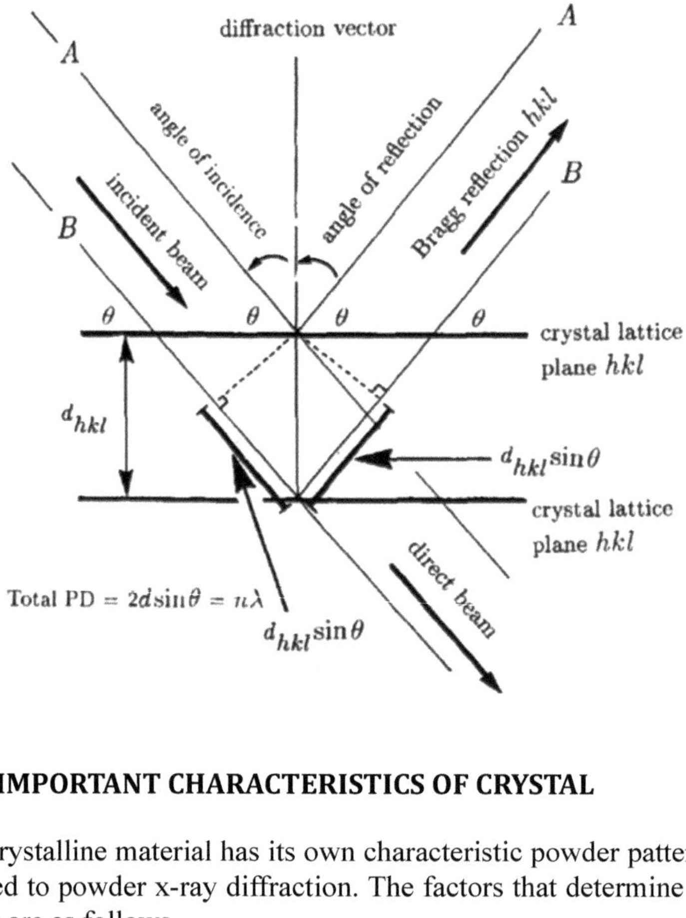

3.10 IMPORTANT CHARACTERISTICS OF CRYSTAL

Every crystalline material has its own characteristic powder pattern when subjected to powder x-ray diffraction. The factors that determine powder patterns are as follows.

a) The size and shape of the unit cell
b) Atomic number and position of the various atoms in the cell

It is possible that two materials may have the same crystal structure but distinct powder patterns.

In case of potassium halides, for example, KF, KCl, and KI, both the positions and intensities of the lines are different in each case though they all have the rock salt structure.

The d-spacing of the lines is shifted because the unit cells are of different size. Intensities of potassium halides are different even though the atomic coordinates are the same for each.[58–60]

The d-spacing and their intensity are characteristic features of powder pattern. The d-spacing should be reproducible from sample to sample and is measured precisely, whereas intensities are more difficult to be measured quantitatively and vary from sample to sample. Thus, a powder pattern of a crystal is its 'fingerprint'.

3.11 SOME IMPORTANT IONIC STRUCTURE

3.11.1 ROCK SALT STRUCTURE (NACL)

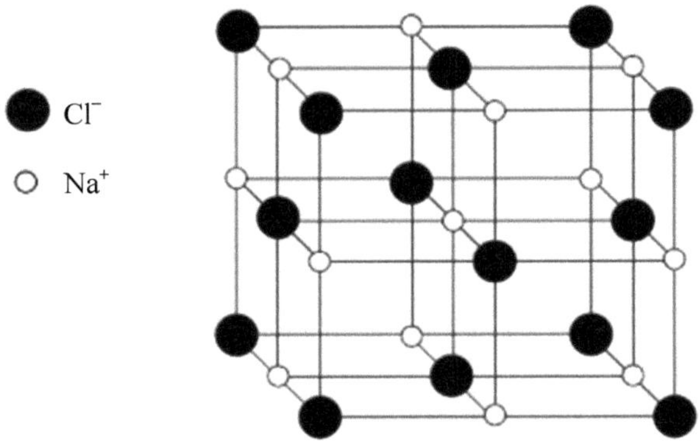

Crystal structure of NaCl

In the ionic structure of NaCl, anion Cl^- is somewhat larger than cation. A closed paced layer of anions with the cations placed in interstitial sites is present. Thus, the structure of NaCl is a face-centered cubic, in which coordination numbers of Na^+ and Cl^- are six having a stoichiometric ratio of 1:1. This is a type of ionic crystal in which all the anions occupy all the octahedral voids and the cations have a face-centered cubic arrangement. Na^+ cation has an atomic radius of 102 pm, whereas Cl^- anion has an atomic radius of 181 pm.

The unit cell of NaCl can be explained as containing four atoms or ions. One corner is shared by eight corners of an ion. So one corner contributes

$\frac{1}{8}$th to the unit cell. There are six faces in a cube which contain one atom in the face.

So the number of atoms in a unit cell $= 8 \times \frac{1}{8}$ atoms $+ 6 \times \frac{1}{8}$ atoms

$$(Z) = 4 \text{ atoms}$$

Each ion in this lattice has six ions as its nearest neighbors and twelve ions as the second nearest neighbor.

There are two types of interstitial sites, tetrahedral and octahedral present in cubic close-packing structure. It is clear that tetrahedral interstitial sites are vacant and octahedral interstitial sites are occupied in rock salt structure. There are a wide range of compounds having similar rock salt structure, for example, CsF, AgF, AgCl, BaO, CaO, and SrS.

LiCl, KCl, and RbCl are some of the alkali metal halides having similar structure to rock salt.

3.11.2 CESIUM CHLORIDE STRUCTURE (CSCL)

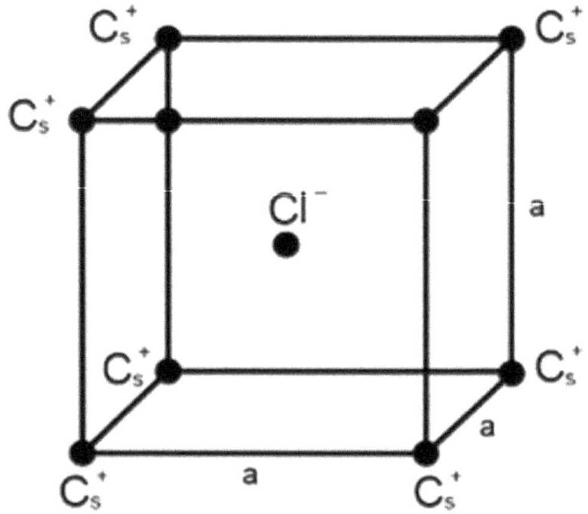

CsCl is a primitive cubic in which eight Cl^- ions are situated at the corners of a cube and a Cs^+ ion remains at the body of the cube. It is not body-centered cubic because there are different ions at the corner and body center positions. The co-ordination numbers of both Cs^+ and Cl^- are eight. The interatomic distance is 0.866 a. In CsCl each Cl has only six Cl^- ions as next nearest neighbors. The CsCl structure cannot be treated as close packed. The compounds showing similarity to CsCl structures are CsCl, CsBr, CsI, CsCN, CuPd, AuZn, etc.

The number of atoms per unit cell in cesium chloride is 2. One atom at the corner contributes $\frac{1}{8}$ to the unit cell and one atom at the body contributes one atom. Hence, the total number of atoms in the unit cell is 2.

When cesium chloride (α-CsCl) is heated to 445°C, this converts into the β-CsCl form with the rock salt structure. The rock salt structure is seen in nanometer thin CsCl films grown on mica.

3.11.3 AX STRUCTURE, FEO

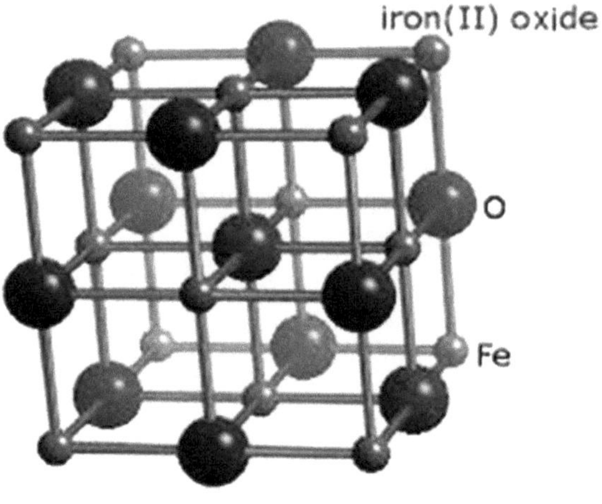

At low temperature magnetic ordering takes place along with the rhombohedral distortion. It is similar to the rock salt structure. A slight distortion of angle from 90° to 90.07° takes place when it is compressed slightly along one threefold axis. Iron atoms are coordinated by oxygen atoms octahedrally. Oxygen atoms are also co-ordinated octahedrally. When oxidation of ferrous ion to ferric ion takes place, the stoichiometry of 1:1 is disturbed. The Fe^{3+} ions occupy tetrahedral positions in the lattice of close packed oxide.

When the crystalline solid is melted, iron atoms are coordinated by four or five atoms only. Stoichiometric ratio of FeO is 4:4, that is, 1:1. In the unit cell of FeO, 4 Fe atoms and 4 O atoms are present. It has also iron deficient nonstoichiometric composition $Fe_{0.84}O$ and $Fe_{0.95}O$.

3.11.4 AX COMPOUNDS WITH DIFFERENT STRUCTURES

The compounds PdO and PtS have d^8 ion configuration. With d^8 configurations the structure is almost a square planer.

Compounds like PbO and SnO show distorted polyhedral because Pb^{+2} and Sn^{+2} exhibit an inert pair effect. When we move down the group, ns^2 electron becomes inert. As a result 2+ oxidation state becomes stable. Thus SnO and PbO show distorted polyhedrals Pb^{2+} and Sn^{2+} have four O^{2-} neighbors to one side giving rise to a square pyramidal structure.

3.11.5 ZINC BLENDE STRUCTURE (ZNS)

ZnS structure is considered the close packed array of sulfide anions with Zn^{2+} cations in tetrahedral sites. The same structure is obtained by the interchange of Zn^{2+} and S^{2-} ions. So the ZnS structure may also be considered a close-paced array of Zn^{2+} ions with S^{2-} ions which occupies tetrahedral sites. Zinc blende has a stoichiometric ratio of 1:1. It maintains a tetrahedral arrangement. The ionic radii of Zn^{2+} and S^{2-} ions are 74 pm and 190 pm respectively. The ratio $\frac{r_+}{r_-} = \frac{74\,pm}{190\,pm} = 0.39$ which suggests a tetrahedral structure. Coordination number of Zinc and Sulfur atoms is 4, that is, each zinc atom is surrounded by four zinc atoms.

3.11.6 DIFFERENCE BETWEEN ZINC BLENDE AND WURTZITE

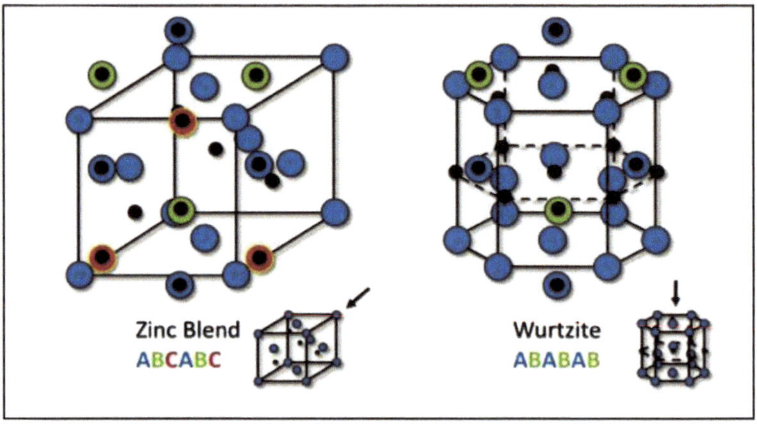

Zinc sulfide has two different crystal structures, zinc blende and wurtzite. Zinc blende is cubic but wurtzite is hexagonal. The two structures are polymorphs of ZnS. From the thermodynamic point of view, zinc blende is more stable in comparison to other forms of ZnS. If temperature is changed, zinc blende becomes wurtzite. The cubic close packed structure of zinc blende is denser than the wurtzite structure. Wurtzite has the hexagonal crystal structure known as hexagonal close packing structure (hcp). There are 12 ions in the corners of each unit cell.

3.11.7 CUBIC ZINC SULFIDE (ZNS) STRUCTURE

The cubic zinc sulfide structure is obtained when Zn atoms are on one fcc lattice and S atoms on the other fcc lattice.

The coordinates of the Zn atoms are 000, $0\frac{1}{2}\frac{1}{2}, \frac{1}{2}0\frac{1}{2}, \frac{1}{2}\frac{1}{2}0$ and those of the S atoms are $\frac{1}{4}\frac{1}{4}\frac{1}{4}, \frac{1}{4}\frac{3}{4}\frac{3}{4}, \frac{3}{4}\frac{1}{4}\frac{3}{4}, \frac{3}{4}\frac{3}{4}\frac{1}{4}$. The lattice is a face-centered cubic.

The diamond structure has an operation of inversion. The inversion operation means moving an atom by the same distance in the opposite direction. The cubic ZnS structure has no inversion of symmetry.

Structures similar to cubic ZnS structure are CuF, SiC, CuCl, GaP, ZnSe, GaAs, AlAs, CdS, InSb, and AgI.

3.11.8 FLUORITE STRUCTURES

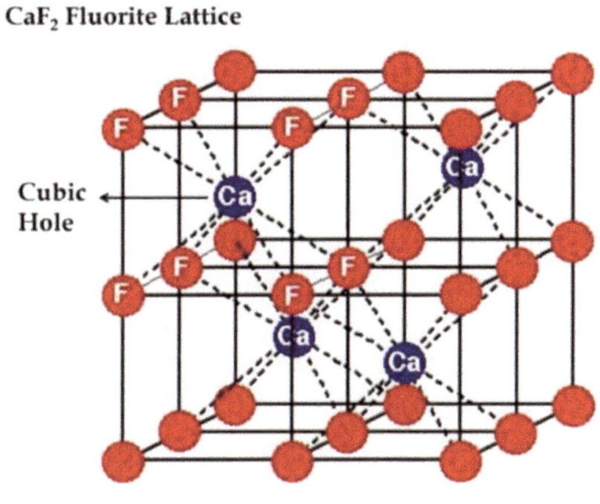

The mineral fluorite CaF_2 possess the structure in which eight tetrahedral interstitial sites are occupied by F^- ions and regular sites are occupied by Ca^{2+} ions a face-centered cubic so the compound having such type of structure is known as a fluorite structure. It is clear that the cations are eight coordinate and the anions are four coordinate. Fluorite structure can also be described as a primitive cubic array of anions. The actual lattice type of the fluorite structure is face-centered cubic.

The density of the fluorite structure of CaF_2 is 3.18 g/cm^3 and the edge length is 5.463 Å.

- When calcium fluoride is in a single molecule, the molecule resonates between a linear and bent shape. Electrons of CaF_2 are contained in the 3d orbitals. The molecule is linear when electrons are in the d_{z^2} orbital and the molecules are bent when they are in the d_{yz} orbitals. The molecule resonates between these two shapes.
- The fluorine-fluorine distance in CaF_2 is 2.73 Å. This shows that the fluorines touch each other. Some of worth mentioning fluorite structure types are $SrCl_2$, $BaCl_2$, CdF_2 and HgF_2.

3.11.9 ANTIFLUORITE STRUCTURE

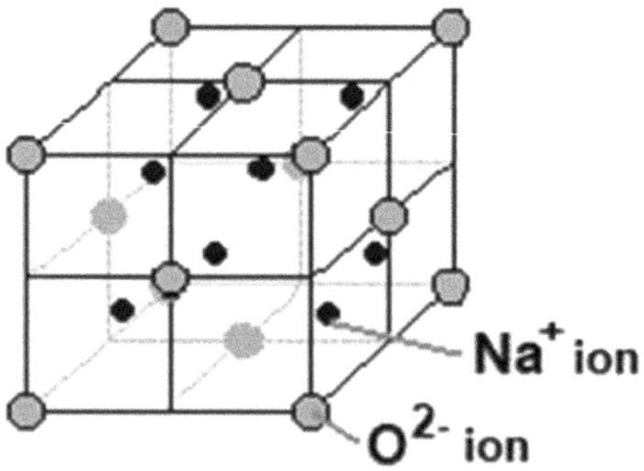

The antifluorite structure has cations in the tetrahedral interstitial site and anions in the regular sites of a face-centered cubic. The antifluorite structure has been shown by a number of alkali metal oxides and chalcogenides.

In the antifluorite structure, anions are eight coordinate and cations are four coordinate. Thus, positions of the cations and anions are reversed to that of the fluorite structure. In the antifluorite structure coordination number is 12 for a face-centered cubic and this unit cell has four atoms only.

Some common examples of antifluorite structure are Li_2O, Na_2O, K_2O, Rb_2O, and Rb_2S.

3.11.10 CADMIUM IODIDE (CDI_2) STRUCTURE

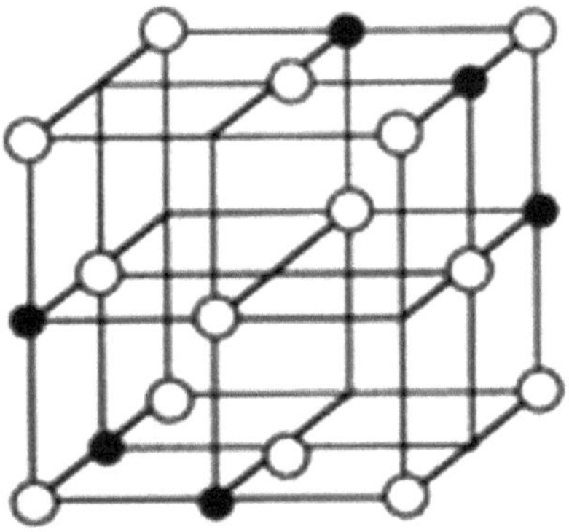

CdI_2 may be explained as the structure in which the iodide anions form a hexagonal close packed arrangement and cadmium ions hold octahedral voids in alternating layers. The CdI_2 structure may be treated similar to rutile, but the occupancy of the octahedral sites is different. Entire layers of octahedral voids get filled which alternate with empty site layers. Rutile is rigid and three dimensional, whereas CdI_2 is a layered material in view of its crystal structure.

The octahedral voids occupied by Cd^{2+} are in between two iodide layers in a hexagonal close packed array. The octahedral sites remain empty to either side of iodide layers .Obviously, hexagonal close packing is represented as –ABABAB—sequence. AB position stands for I^- and Cs^{2+} ions occupy octahedral site. In a more simple way, it can be explained as Cd^{2+} ions sandwiched between two layers of I^- ions. The binding forces between them are weak Vander Waals bonds.

CdI_2 may be considered an infinite sandwich molecule in which the weak Vander Waals force exists between adjacent molecules but Cd-I bonds within a layer are strong. This is why cadmium iodide is predominatingly ionic having some covalent character too.

A few compounds show similar characteristic and structure, for example, hydroxides of di cations $M(OH)_2$, sulfides of tetracations MS_2, $CdBr_2$.

The density of CdI_2 is 5.67 g/cm^3 and soluble in water, ethanol, acetone, and ether. Other halides of cadmium are similar to structure of cadmium iodide. In case of cadmium chloride the only difference lies in the anion packing. Cl^- ions are cubic close packed in $CdCl_2$ but in $CdCl_{2,}$ I- ions are hexagonal close packed.

3.11.11 DIFFERENCE BETWEEN THE STRUCTURES OF CDI2 AND CDCL2

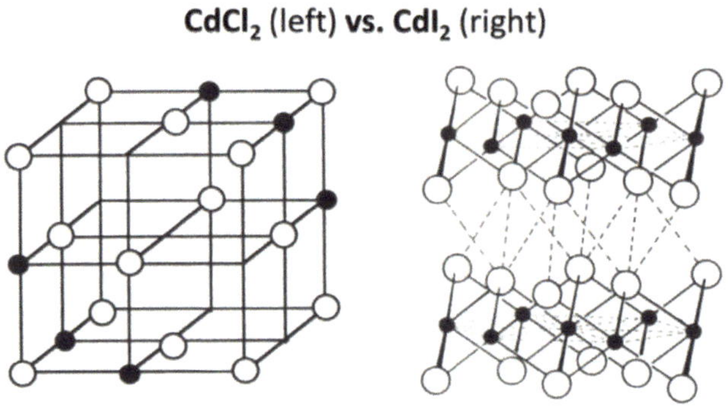

The unit cell in a cadmium chloride structure is hexagonal. The base of the hexagonal cell is similar to that in CdI_2 but the difference lies in the length of c-axis. It has been established that the c-axis of $CdCl_2$ is three times longer than $CdI_{2.}$ This may be explained due to Cd^{2+} positions and staggering along the c-axis.

Compounds showing similarity with $CdCl_2$ structure are $CoCl_{2,}$ $FeCl_{2,}$ $MgCl_{2,}$ and $MnCl_{2.}$

3.11.12 STRUCTURE OF CS_2O

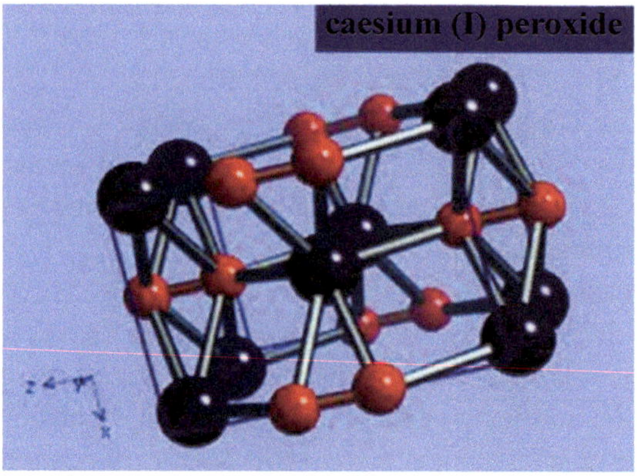

The structure of Cs_2O resembles anti-$CdCl_2$ structure. Cs^+ ions constitute cubic close packed layers, whereas oxide ions occupy the octahedral sites between alternate pairs of cesium layers. In the Cs_2O structure Cs^+ ions are not surrounded by oxide ions. Cs^+ ion has only three oxide ions as neighbors. Bonding is between cesium ions in adjacent layers. The ratio of coordination numbers of Cs^+ and O^{2-} is 1:2 because the ionic radius of Cs^+ is larger than the ionic radius of O^{2-}. The maximum coordination number of oxygen by cesium can be six only. This says C.N. of Cs^+ is three.

3.11.13 RUTILE STRUCTURE (TIO_2)

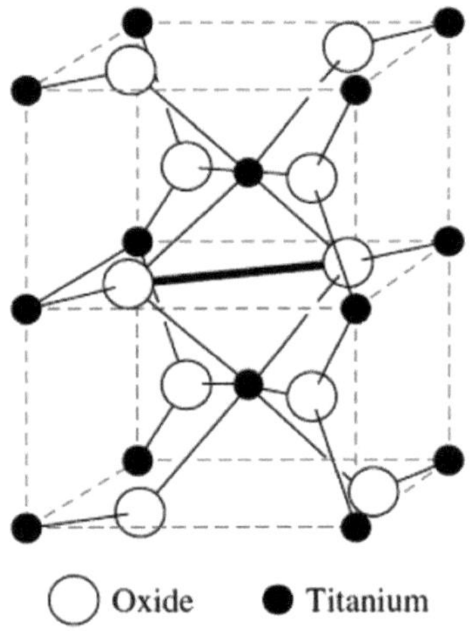

In the rutile structure TiO_2, the titanium atom is fixed at the corner and body center. The crystal system is tetragonal. Rutile is a mineral mainly of titanium dioxide. Oxygen atoms form the corner of TiO_6 octahedra. Each corner oxygen is shared between three octahedra that are linked by sharing edges and corners. Thus, a three-dimensional network is formed. It may also be considered a tetragonal packed array or a distorted hexagonal paced array. One-half of the octahedral sites get occupied by Ti^{4+}. The TiO_6 octahedra share edges parallel to the tetragonal C-axis. Thus, it is formed by quasi-hcp packing of anions. Oxides of some tetravalent metal ions and fluorides of small divalent metal ions show a rutile structure. The rutile structure may be considered an ionic structure. In the tetragonal unit cell of rutile unit cell parameters $a = b = 4.584$ Å $c = 2.953$. The coordination number of titanium cations is 6. Titanium cation is surrounded by an octahedron of 6 oxygen atoms. The oxygen anions have a coordination number 3.

3.11.14 PEROVSKITE STRUCTURE

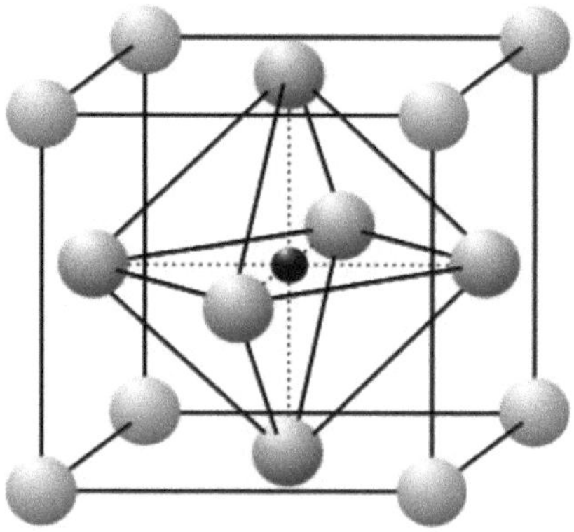

The general formula of the perovskite-related structure is ABO_3, $BaTiO_3$, $NaNbO_3$, and $KTaO_3$ have distorted perovskite structure at ordinary temperature and show cubic structure at high temperature. Perovskites may give rise to nonstoichiometric compositions, for example, $A_{(1-x)}$ BX_3 and $AB\ X_{3-x}$ for example A_xWO_3.

Perovskites are of two types

i) $A^{II}\ B^{IV}\ O_3$
   $A^{I}\ B^{V}\ O_3$

Some of perovskites are ferroelectric $GdFeO_3$ that is orthorhombic and $LaAlO_3$ that is rhombohedral type perovskites.

3.12 SPINELS

One of the most common examples of spinel is $MgAl_2O_4$. The structure of spinel has a cubic close-packed array of O^{2-} in which magnesium ions occupy $\frac{1}{8}$ of the tetrahedral A sites and the aluminum cations occupy half the octahedral B sites. So normal spinels can be represented as $(A)_t\ [B_2]_oO_4$. Spinel of type $(B)_t\ [AB]_o\ O_4$ is called inverse spinel in which half of the B ions are in the tetrahedral site and the other half along with A ions occupy octahedral sites. Magnetite Fe_3O_4 has the inverse spinel structure described as $Fe_t^{3+} [Fe^{2+} Fe^{3+}]_o O_4$.

3.13 NONCRYSTALLINE SOLID

Noncrystalline solid, generally known as amorphous solid, has no precise structure because positions of the atoms are only short range. The exact number of nearest neighbors is difficult to be known in amorphous solid. The unit cell of the lattice is not regular. From X-ray diffraction of amorphous solid, it is clear that a glass or amorphous solid has broad diffuse rings that are entirely different from a large number of sharp rings in crystalline solid. So the unit cell cannot repeat regularly in three dimensions in glass. Obviously, the number of nearest neighbors is not the same in amorphous solids. From the glass diffraction pattern only the radiation distribution function is determined directly.

3.13.1 STRUCTURE OF VITREOUS SILICA, SiO_2

With the method of fluorescence excitation, the structure of vitreous silica has been explained. Each of Si atoms is tetrahedrally surrounded by 4 oxygen atoms. The Si-O distance is close to 1.62 Å. The first peak is at 1.62 Å which is very similar to the Si-O distance in crystalline silicates. The tetrahedron geometry predicts that the O-O distance is 2.65 Å. The tetrahedral groups have different orientations with respect to neighboring group about the Si-O-Si bond. The unit cell in the structure does not repeat itself at regular intervals in space. As a result the matter is no more crystalline. There is an important distinction between the vitreous and the crystalline forms of silica. It can be said that glass has a random network model.

The random structure of glass is obtained by cooling the liquid below the freezing point. Obviously, a glass has the elastic properties like an isotropic solid. A liquid on cooling changes to glass when the viscosity attains a value of 10^{13} poise. At glass transition temperature T_g, we have liquid and below T_g glass exists. But this transition of phase has no thermodynamic nature. Few liquids change to glass followed by fast cooling in the bulk. Vitreous silica is prepared by melting quartz or cristobalite.

When glass is crystallized, either quartz or cristobalite may appear. Molten SiO_2 is viscous even at 1800°C. It cannot be ensured complete melting and homogenization even at this temperature. This model is known as microcrystalline model.

3.13.2 SILICATE GLASSES

The type of silicate glasses possesses regular network of SiO_4 tetrahedra in a random arrangement through the oxygen atoms at the apexes. It may also be described as covalently bonded glasses. In crystalline silicates, for example, $Na_2O.2SiO_2$, the silicate anions are usually infinite two-dimensional sheets. An open three-dimensional structure exists where cations Na^+ occupy large holes in the network. X-ray diffraction shows that the distribution of Na^+ in the framework is random. The tetrahedral units of SiO_4 are shared by one corner oxygen ion which give rise to a short-range structure of silicate glasses. The glass-forming oxides may substitute for silica and consequently a random three-dimensional network structure is retained. Lead silicate glasses may be prepared by substitution of few SiO_2 units by PbO. Thus PbO is a glass-forming oxide. B_2O_3-SiO_2 glasses may also be prepared by substitution in any range of composition between SiO_2 and B_2O_3.

3.13.3 BORATE GLASSES

B_2O_3 is present in borosilicate glasses such as pyrex. B_2O_3 glass contains BO_3 triangles and BO_4 tetrahedra. An important constituent of vitreous B_2O_3 is the boroxyl group which has planar six-membered ring having alternate boron and oxygen atoms. Molten B_2O_3 is more fluid than molten SiO_2. When alkali oxide is added to vitreous B_2O_3, for example, Na_2O-B_2O_3, the viscosity of the melt increases with increasing Na_2O. Alkali silicates become more fluid, their coefficient of thermal expansion increases with alkali content. It has been observed by Bray using B^{11}NMR spectroscopy that coordination number of boron changes from three to four on addition of alkali oxide to B_2O_3.When 30% alkali oxide is added, 40% of the borons convert into tetrahedral coordination independent of the properties of the alkali.

3.13.4 PHASE SEPARATION IN GLASSES

Glasses form a homogeneous single phase. When cooling takes place in a homogeneous melt, single phase of glass is obtained. Liquids undercool to form glass without any crystallization. The appearance of a glass containing immiscibility texture depends on the wavelength of visible light. The liquid having a composition of 30% CaO and 70% SiO_2 gives opaque glass on cooling. When refractive index between the glassy phase differs, the glass will be opaque. Lithium silicate glass having 25% lithium oxide appears blue in reflected light and orange in transmitted light. Scattering of some visible light takes place at the internal boundaries. If the two glassy phases have the same refractive index, scattering of light does not take place at the internal boundary and as a result the glass appears transparent. Glass having 20% Na_2O and 80% SiO_2 looks like a transparent glass.

3.13.5 THERMODYNAMIC CRITERIA

Gibbs free energy change, enthalpy change, and entropy change are the thermodynamic parameters that decide the immiscibility of two liquids A and B. The expression thermodynamic relation as $G = H - TS$. Formation of a single-phase homogeneous liquid is accompanied by entropy increase. Interactions between A-B, A-A, and B-B are also important to decide the immiscibility. When A and B are mixed, the enthalpy change may be either negative or zero. The attractive interaction between A and B is greater than or equal to the attractive interactions between A-A and B-B.

$\Delta H \leq 0$, If $\Delta H > T\Delta S$, immiscibility occurs. It may be concluded that immiscibility occurs at low temperature. When temperature is increased, miscibility increases. When $\Delta H < T\Delta S$, complete miscibility occurs above the consolute temperature T_c.

3.13.6 MECHANISM OF PHASE SEPARATION

When a homogeneous liquid enters immiscibility dome, separation of the homogeneous liquid takes place in two liquid phases. Two mechanisms have been proposed. (I) Nucleation and growth. (II) Spinodal decomposition. In both the mechanisms, free energy change with varying composition takes place in a different manner. But if fluctuations in composition take place in a liquid mixture, an increase in free energy takes place. But the fluctuation is unstable and the liquid gains homogeneity. Nucleation and growth takes place at a slow rate. Nucleation may be shown by the graph of activation energy. Fluctuations in composition in liquid mixture arise from random thermal motions of atoms in the liquid. In spinodal decomposition, the spontaneous phase separation takes place very fast. A large decrease in enthalpy takes place on phase separation leading to an overall decrease in free energy. The mechanisms have been proposed by Cahn and Charles. Some of the basic differences between nucleation and growth and spinodal decomposition are as follows: (i) In nucleation and growth mechanism, a sharp boundary exists between the glassy phases whereas in spinodal decomposition, phases are diffuse in the beginning but become sharp with time. (ii) In nucleation and growth mechanism, composition of second phase remains unchanged with time whereas compositions of both phases vary with time till equilibrium.

3.13.7 VISCOSITY OF GLASS

Viscosity is an important parameter that decides the manufacture of glass. If the viscosity of molten mixture is high at a certain temperature, long time is required for the crystalline material to dissolve. The gas bubbles are removed for achieving good quality product and so viscosity is decreased.

After melting, the material is cooled to form a glass. In the region of glass transition, it is heated further for annealing to remove mechanical stress and giving a final shape to the glass.

The viscosity of a glass at different stages is

Annealing point $10^{13.4}$ P

Glass transition 10^{14}–10^{15} P

P is the unit of viscosity in poise (P) which has the unit (gm.cm^{-1}.sec^{-1}).

Arrhenius equation is followed by viscosity of simple liquids

$$\eta = \eta_o \exp^{\frac{\theta}{RT}}$$

where θ is the activation energy for viscous flow.

$$\log \eta = \log \eta_o + \frac{\theta}{RT}$$

When η is plotted against $\frac{1}{T}$, slight deviation from straight line is obtained. Fulcher proposed some correction to explain the validity of data.

$$\eta = \eta_o \exp^{\frac{B}{T-T_o}}$$

where B is constant, T_o is the lower limiting ideal temperature at which the viscosity becomes very large.

If $T >> T$, the Fulcher equation may fit the data but for $T \rightarrow T_o$, deviations may occur.

3.13.8 COMMERCIAL GLASSES

Commercial glasses are used in windows, bottles, lamps, and other purposes as pure silica glass has a high transition temperature. When Na_2O is added to SiO_2, the melting temperature drops from 1700°C to 800°C. The homogeneous melt is obtained in lesser time. Further addition of CaO, MgO, and Al_2O_3 to Na_2O-SiO_2 decreases the melting temperature to lower range. Chemical durability increases as a result. Commercial glasses are, therefore, multicomponent material.

Annealing of glasses is necessary for uniformity and stabilization. High-quality optical glasses are uniform and have constant refractive index throughout the phase. Annealing also helps relieve mechanical stress. Generally annealing time of a few hours is required. Sodium borosilicate glass of 10% Na_2O, 30% B_2O_3, and 60% SiO_2 composition is also useful for practical purposes.

Pyrex glass is also very useful which is made from 4% Na_2O, 16% B_2O_3, and 80% SiO_2.

3.14 AMORPHOUS SEMICONDUCTORS

Amorphous semiconductors are prepared from thin films by supercooling the melt. Sulfur and selenium give viscous liquids on melting and glasses are formed when cooled. The liquids contain rings and chains of sulfur and selenium atoms. Such amorphous semiconductors have important commercial uses. Selenium is used in photocopying machines.

It is clear that glasses have no regular crystalline order.

Conductivity of amorphous semiconductors can be explained by both electrons and holes. The carriers are scattered due to a disordered structure. Amorphous materials resemble intrinsic semiconductor behavior, in which Fermi level lies near to the center of the gap. There are two important classes of amorphous semiconductors: (i) Tetrahedrally bonded amorphous solids, for example, silicon and germanium, (ii) Chalcogenide glasses, for example, sulfur, selenium, or tellurium as one component.

Tetrahedrally bonded amorphous solids are similar to those of crystalline forms. They can be doped with small amount of chemical impurities to enhance the carriers.

Chalcogenide glasses are not sensitive to chemical impurities. Amorphous hydrogenated silicon is used for solar cells.

Hydrogen has been introduced into amorphous silicon to remove unwanted structural defects. Preparation of pure amorphous silicon is difficult.

3.14.1 AMORPHOUS FERROMAGNETS

When molten alloy is quenched rapidly, amorphous metallic alloys are formed. Amorphous materials have isotropic properties so they have low coercivities, low hysteresis losses, and high permeabilities. The electrical resistivity is high due to randomness in alloys.

The transition metal–metalloid alloys are magnetic amorphous alloys. Transition metal percent is 80% and metalloid component percent is 20%. Commonly used transition metals are Fe, Co, Ni, whereas metalloid components are B, C, Si, P, and Al.

The presence of metalloids as a component lowers the melting point. The alloy is quenched through the glass transition temperature and as a result amorphous phase is stabilized.

Metglas 2605 has 80% Fe and 20% B. Its glass transition temperature is 441°C.

Properties of amorphous semiconductors:

i) Amorphous semiconductors have wide composition ranges. They have no stoichiometric composition.
ii) Generally, doping does not take place in amorphous semiconductors.
iii) Chalcogenide-based glasses have conductivities in the range 10^{-3} to 10^{-13} $ohm^{-1}.cm^{-1}$.

Most germanium atoms are tetrahedrally coordinated in amorphous germanium. Few germanium atoms are three coordinated only.

The possibility of n-type conduction by doping of amorphous germanium with arsenic is minimum. Arsenic dopant atoms do not provide a free electron when three-coordinate sites are occupied by arsenic.

3.15 METALLIC GLASSES

Liquid metals do not supercool to form glasses. Rapid quenching techniques are required to prepare glassy metals. Liquid droplets are projected with high velocity onto a cold surface so that liquid is cooled within fraction of a second. Thin ribbon of glass is prepared by roller quenching in which the melt is propelled onto a cooled rotating drum.

Liquid selenium has covalent selenium linkages. It forms a glass but more metallic bonding restricts liquid tellurium to change to glass. Ionic bonding between Na^+ and Cl^- ions restricts NaCl to form glass.

3.15.1 PROPERTIES

i) Glassy metals are stronger than crystalline metals. Many pure metals are soft due to the presence of dislocations, but glassy metals are usually hard. Dislocations move readily in crystalline structure from one region to other. But the dislocations, if any, are not free to move in glassy metals. Glassy metals show high plasticity.
ii) Glassy metals are resistant to corrosion. Glassy metals have no grain boundaries or dislocations. This is why glassy metals are comparatively less reactive.
iii) Cobalt and iron-containing glassy metals have low coercivity and can be magnetized easily.

3.16 GLASS-CERAMICS

Glass-ceramics possess properties of both glass and ceramics. The mechanical strength of glass ceramic is higher than the glasses. When heat treatment is regulated, nucleation and growth of crystalline phase occur within the glass. Sometimes, crystallization is complete and sometimes residual glass phase may be present due to incomplete crystallization.

Nucleation for glass ceramics may be done by the following methods:

i) Colloidal particles of Cu, Ag, and Au are dispersed in the melt. The metal particles do not dissolve completely, but they act as nucleation sites.
ii) The oxides such as TiO_2, P_2O_5, and ZrO_2 are soluble in the melt at high temperature, but they get precipitated by annealing at low temperature.

The crystal size in the glass-ceramic is in the range 0.1–1 μm. Thus, the overall steps can be summarized as the preparation of glass by melting the component oxides. The melt is cast into a desired shape and cooled. Li_2O-SiO_2 is one of the examples of the glass-ceramic system.

3.16.1 PROPERTIES

i) Glass-ceramics is resistant to abrasion and stronger than general glass.
ii) Glass-ceramics withstands changes in temperature without any damage. So glass-ceramics is resistant to thermal shock.
iii) Glass-ceramics behaves as insulators.
iv) The transparent or opaque nature of glass ceramics may be explained due to difference in refractive index between crystal and residual glass.
v) Glass-ceramics has no porosity.

Uses: Glass-ceramics is used in cooking ware and oven tops. They are also used as bearings due to high temperature stability and thermal shock resistance; they are used as heat-protectors in rockets. Glass-ceramics is also used in electronic industry.

3.17 OPTICAL FIBRES

The optical fibers contain a thin core of high-refractive index glass. Fibers of silica-based light guides carry data and information on the surface of earth. Light beams bounce many times on the walls, that is, total internal reflection takes place. The telephone calls and emails can be transmitted in a fraction of seconds. This technology is used to transmit the calls at the speed of light. Information is coded in a beam of light down a glass. Fiber optics cable is composed of very thin strands of glass. When light beams travel down the cable, they emerge at the other end where information is gathered from pulses of light. The middle part of the cable is core and outside the core another glass exists called cladding. Cladding and total internal reflection of light are responsible for keeping the light inside the pipe. Thus digital data carried by light is recovered at the end.

Exercise 1

For solid, oxide ions in ccp $\frac{1}{6}$ th of the tetrahedral voids get occupied by the cations A whereas 1/3rd of the octahedral voids get occupied by cations B. Find out the formula of the compound?

Solution

Let number of O^{2-} ions be 90.

Octahedral voids = 90

Octrahedral voids = 90

Tetrahedral voids = 180

Number of cations A = $\frac{1}{6} \times 180 = 30$

Number of cations B = $\frac{1}{3} \times 90 = 30$

A: B: O^{2-} = 30: 30: 90

= 1: 1: 3

The desired formula is ABO_3

Exercise 2

Oxide ions have hexagonal close packing and aluminum ions occupy 2/3rd of octahedral voids in corundum. Find the formula of corundum.

Solution

Let number of oxide ions be 90

Total no. of octahedral voids = 90

Number of $Al^{3+} = \frac{2}{3} \times 90 = 60$

Ratio of $Al^{3+}: O^{2-} = 60: 90$

$= 2:3$

The desired Formula is Al_2O_3

Exercise 3 Find the percentage of iron as Fe^{3+} in $Fe_{0.93}O$.

Solution: The total charge of Fe^{2+} and Fe^{3+} should be 2

Let percentage of $Fe^{3+} = x$

So $3 \times x + 2\,(0.93 - x) = 2$

$$x = 0.14$$

$$\% \text{ of } Fe^{3+} = \frac{0.14}{0.93} \times 100 = 15\%$$

KEYWORDS

- **unit cell**
- **lattice plane**
- **powder diffractometer**
- **rutile structure**
- **spinel**
- **silicate glass**

CHAPTER 4

Space Groups and Point Group

ABSTRACT

This chapter gives us a brief knowledge of point groups and space groups. There are conventions to label axes in a space group and graphical representation of symmetry elements have also been dealt with. Point groups, Bravais lattice and different possible lattice types give rise to 230 space groups. Space groups are formed by adding elements of translation to the point groups. The space symmetry element, screw axis and glide planes are derived from point symmetry elements, rotation axes and mirror planes by adding translation.

4.1 POINT GROUPS

The elements of point symmetry in crystals are rotation axes, inversion axes, and the mirror plane m. These symmetry elements alone or in various combinations give 32 possible crystallographic point groups. The symbols for different point groups are as follows.

TABLE 4.1 Symmetry Elements.

Symmetry element	Written symbol	Graphical symbol
Rotation	2	

Solid-State Chemistry: A Modern Approach. Ashok Kumar Jha, PhD (Auth.)

TABLE 4.1 *(Continued)*

Symmetry element	Written symbol	Graphical symbol
	3	
	4	
	6	
Inversion	$\bar{2} \equiv (m)$	–
	$\bar{3}\ (\equiv 3+T)$	
Axis	$\bar{4}$	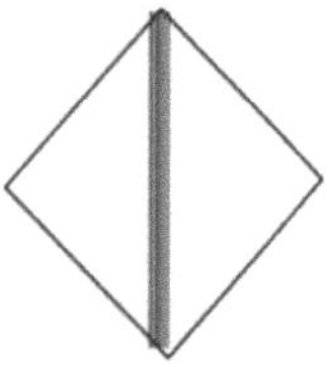
	$\bar{6}\ (= \frac{3}{m})$	
Mirror plane	M	–

Graphical Representation of Symmetry Elements

- Proper rotations depicted as symbols with the number of vertices which corresponds to n
- Screw axes have same symbol, but have "tails"
- Enantiomeric pairs of screw axes (*e.g.* 6_1 and 6_5) are mirror images of each other

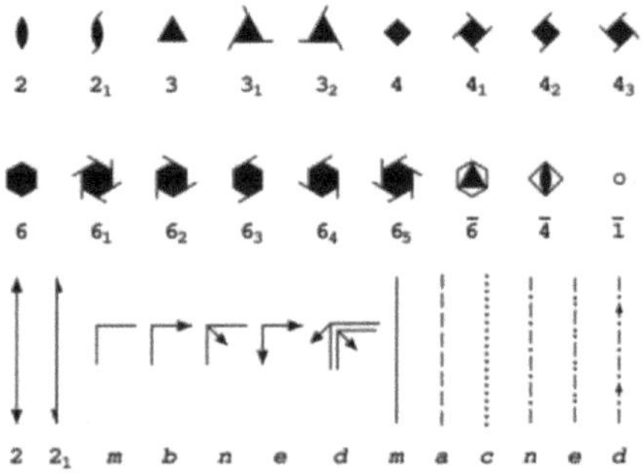

TABLE 4.2 Symbol of Symmetry Elements.

Symmetry Element	Graphical Symbol	Translation	Symbol
Identity	None	None	1
2-fold ⊥ page		None	2
2-fold in page		None	2
2 sub 1 ⊥ page		1/2	2_1
2-sub 1 in page		1/2	2_1
3-fold	▲	None	3
3 sub 1		1/3	3_1
3 sub 2		2/3	3_2
4-fold	■	None	4
4 sub 1		1/4	4_1
4 sub 2		1/2	4_2
4 sub 3		3/4	4_3
6-fold	●	None	6
6 sub 1		1/6	6_1
6 sub 2		1/3	6_2
6 sub 3		1/2	6_3

TABLE 4.3 Crystal System and Point Group.

Crystal system	Point group
Triclinic	$1, \bar{1}$
Monoclinic	2, m, 2/ m
Orthorhombic	222, mm^3, mmm
Tetragonal	$4, \bar{4}\ {}^{4}/_{m}$, 422, 4mm, $\bar{4}2m$, 4/mmm
Trigonal	3,3_,32,3m,3_m
Hexagonal	$6, \bar{6}, \frac{6}{m}$, 622, 6mm, $\bar{6}$ m2, $\frac{6}{mmm}$
Cubic	23, m3, 432, $\bar{4}3m$, m3m

4.2 REPRESENTATION OF POINT GROUPS

The point groups are represented by a circle in which one of the axes is perpendicular to the plane of the circle usually passing through its center. For orthorhombic point group:

222	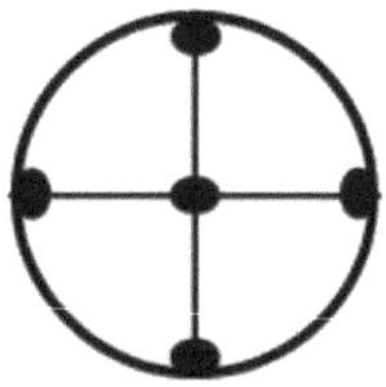
mm2	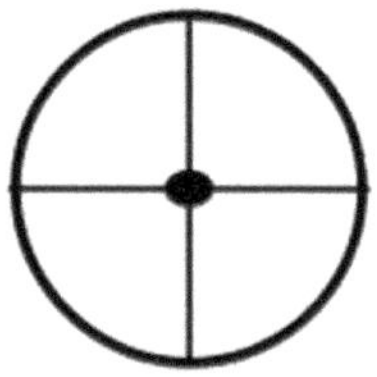
mmm	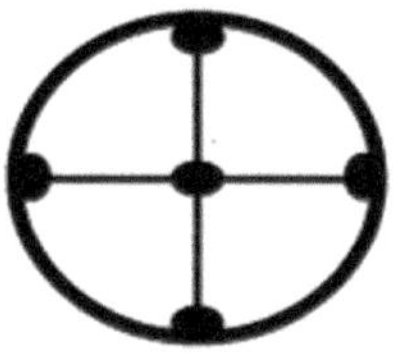

222: The axis perpendicular to the plane of the paper is represented by the symbol in the center of the circle. The axis that lies horizontally and vertically in the plane of the paper is represented by the two respective pairs of symbols on the circumference of the circle.

The presence of three mutually perpendicular twofold axes will give four equivalent positions.

The sequence of steps in point group 222 for deriving the equivalent positions is given below.

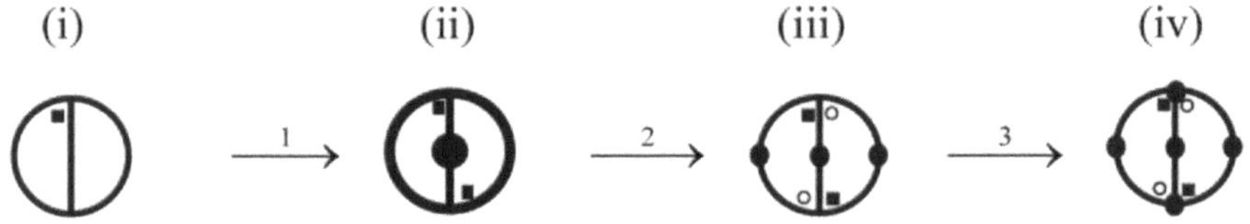

The thin vertical line that bisects a circle is a construction line. When a twofold axis is added perpendicular to the plane of the paper, (ii) is obtained. When a second twofold axis is added in the horizontal direction, (iii) is obtained. Comparison of (iii) and (iv) indicates that both (iii) and (iv) are identical, that is, four positions are already present in (iii). Thus addition of third twofold, in the plane of the paper and lying vertical, does not give any equivalent position. The four positions are already present. Open circles show that new positions generated by twofold axis are below the plane of the paper.

In step 1, a twofold axis perpendicular to the plane of the paper is added. In step 2, a second twofold axis is in the plane of the paper, that is, the second twofold axis is horizontal. In step 3, a third twofold axis lies vertically in the plane of the paper.

mm^2: In the mm^2 orthorhombic point group there are two mirror planes at right angles to each other and a twofold axis passes along the line of intersection of the mirror planes.

I  II 

The mirror planes are represented by thick lines when they lie horizontal and vertical. Four equivalent positions are represented by points 1, 2, 3, and 4 in (ii). If 1 is the starting point, the equivalent position 3 is obtained after a twofold axis operation. If the vertical mirror plane is there, position 2 from 1 and position 4 from 3 are obtained. The horizontal plane does not create any new position but explains equivalent positions 1 and 4 and 2 and 3. All the positions mentioned in II, for example, 1, 2, 3, 4 by points are equivalent.

mmm: This point group mmm indicates that three mirror planes are perpendicular to each other. As a result of this, three mutually perpendicular twofold axes are generated. The thick circle in I indicates the presence of a mirror plane in the plane of the paper.

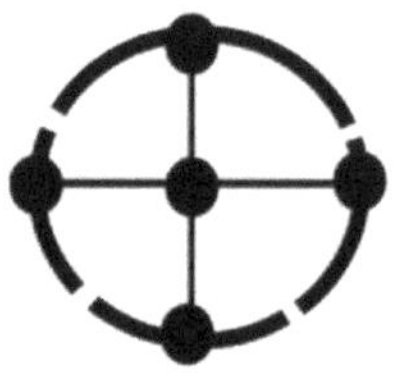

The operation of any two mirror planes gives four positions in a plane given below.

The operation of the third mirror plane which lies in the plane of the paper gives the eight equivalent positions given below.

These eight equivalent positions may be thought of four at the same height above the plane of the paper and four at the same height below the plane.

In addition to this, out of there orthorhombic point groups, mmm point group possesses a center of symmetry. Inversion through this center gives equivalent positions 1 and 2.

Space groups are made up from

- Lattice symmetry (translational)
- Point symmetry (not translational)
- Glide and/or screw axes (some translational component)

There are **230** space groups

7 Crystal System → **14 Bravais Lattices** → **32 Point Group** → **230 Space Groups**

Space group is a fundamental character of crystal. There are 230 possible space groups to characterize the crystals. The space symmetry involves a combination of rotation with translation or mirror plane reflection with translation.

The screw axis combines translation and rotation. The symbol for a screw axis x_Y means transition by the fraction $\frac{y}{x}$ of the unit cell edge in the direction of the screw axis is followed by a rotation of $\frac{1}{x} \times 360°$ about the screen axis. For example, $a4_2$ axis means translation by $\frac{a}{2}$ parallel to a and rotation by 90°.

The glide plane combines translation and reflection. Translation may be parallel to any of the unit cell axis a, b, c and n glide planes all have translation step of half the unit cell in that direction.

For the axial glide plane a, b, and c, both the direction of translation and the reflection plane may be perpendicular to ac plane or b.

Thus, 32 possible point groups and the 14 Bravais lattice give rise to 230 possible space groups. The crystal structure belongs to any of the possible space groups.

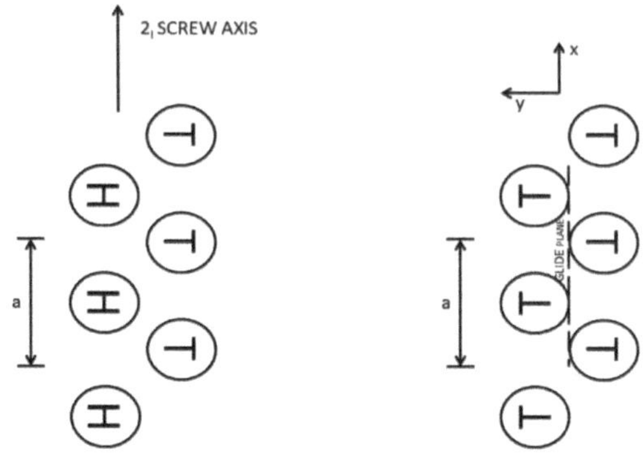

FIGURE 4.1 Arrangement of coins with heads (H) and tails (T), showing (a) a 2_1 screw axis parallel to a and (b) a glide plane perpendicular to b.

It is clear that space groups are formed by adding elements of translation to the point groups. The space symmetry elements, screw axis, and glide planes are derived from point symmetry elements, rotation axes, and mirror planes by the addition of a translation step in between each operation of rotation or reflection.

The written symbol of a space group is a list of between two and four characters. The first character is always a capital letter corresponding to the lattice type—P, I, A etc. The remaining characters correspond to some of the symmetry elements present.

Space groups are drawn as parallelograms with the plane of the parallelogram corresponding to the *x y* plane of the unit cell. For example, the origin is taken as the top left hand corner with *y* horizontal, *x* vertical, and the positive *z* direction pointing up out of the plane of the paper.

For each space group, two parallelograms are used, one on the left hand side to give the equivalent positions and one on the right hand side.

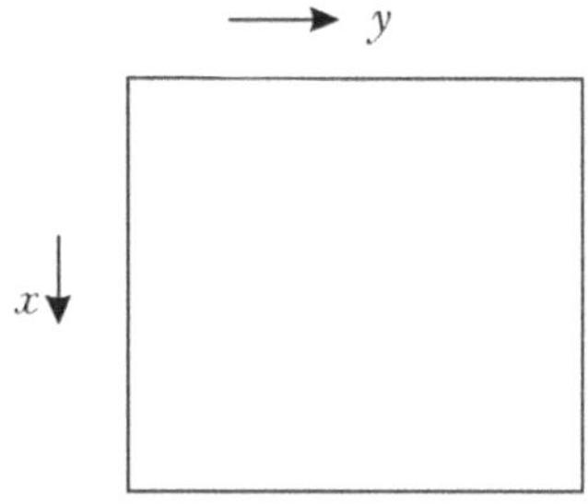

FIGURE 4.2 Convention used to label axes in a space group.

P, F, I, and C designate the lattice type. P stands for primitive lattice containing lattice points at the corners of the unit cell, F face-centered lattice, I body-centered lattice.

During discussion of space groups, a set of spatial coordinates (*x*, y, z) is assigned to every point in the lattice. Shubnikov suggested that a fourth coordinate S is added referring to the spin of a particle, one spin up and the other spin down. A new antisymmetry operation is introduced to reverse the value of S from spin-up to spin-down. If this operation is taken into account, the point groups and space groups become 122 and 1651 respectively. Cracknel referred these groups as magnetic groups or colored groups.

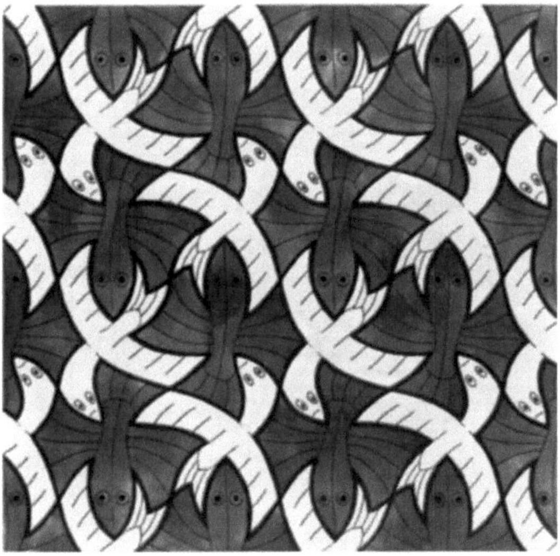

FIGURE 4.3 Only translation with twofold space group.

KEYWORDS

- **point groups**
- **space groups**
- **mirror planes**
- **graphical symbol**
- **translation**

CHAPTER 5

Superconductivity

ABSTRACT

High temperature superconductors have been discovered. If the cooled superconducting sample is placed in a uniform magnetic field, the magnetic flux lines do not penetrate the sample. The basic principles such as Meissner effect, specific heat, isotope effect, etc have been dealt with in detail. Superconductors with Bi, Tl and La have also been discussed. Superconductors without Cu and lanthanides have also been studied. The mechanism of superconductors has been explained by BCS theory. Organic superconductors of TTF – TCNQ type and fullerenes such as K_3C_{60} have also been dealt with. Superconductors have applications such as powerful electromagnet, MRI, frictionless train.

5.1 INTRODUCTION

In 1911, Kammerlingh discovered that the mercury sample lost electrical resistance below 3–7K. At this temperature mercury exhibited the property of zero resistivity below a temperature called the critical temperature and a large number of materials, for example, mixed oxide superconductors and several other superconductors have been discovered since then. Until 1986 some of these superconductors of metal oxides are $BaPb_{1-x}Bi_xO_3$, $Li_{1+x}Ti_{2-x}O_4$ showing property of zero resistivity of around 13K. The highest transition temperature for a superconductor, for example, Hg Ba_2Ca_2 Cu O $_{8+}\partial$ is 155K. Some cuprate superconductors have also been discovered, for example, $La_{2-x}A_x$ CuO_4 (A = alkaline earth), $LnBaCu_3O_{8+}\partial$ (Ln = Y or rare earth other than Ce, Pr, and Tb) referred to as 123 type, Ln $Ba_2Cu_40_8$ (124), and Ln_2 $Ba_4Cu_7O_{15}$ (247).

Solid-State Chemistry: A Modern Approach. Ashok Kumar Jha, PhD (Auth.)

5.2 BASIC PRINCIPLE

Infinite conductivity: When a superconducting substance is cooled below its critical temperature, the resistivity of the sample becomes zero. The conductivity σ becomes infinite. *X*-ray structure analysis shows that there is no change in the crystal structure at this temperature. The relation between current and electric field is as given below $J = \sigma E$

If current density is finite, the electric filed E inside a superconductor must be zero.

5.2.1 MEISSNER EFFECT

If the cooled superconducting sample is placed in a uniform magnetic field, the magnetic flux lines do not penetrate the sample. Since the sample is a superconductor, this will keep magnetic flux out in the absence of any resistance. Thus when such a sample is cooled to a temperature below transition temperature, the magnetic flux inside the sample will be expelled.

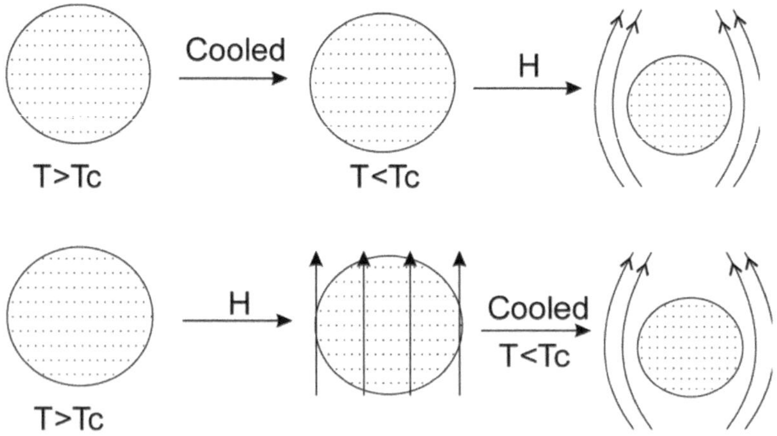

The effect showing this is known as Meissner effect.

In superconductors, the magnetic flux does not penetrate into the sample as long as the applied magnetic field is lower than the critical field He (T). The magnetic field of induction B is related to the magnetization through

$$B = H + 4\pi I$$

A superconductor is characterized by B = 0. Thus when the applied field is less than the critical field, the sample behaves like a diamagnetic. The magnetization becomes zero above the critical field H_c and the sample shows para magnetic behaviors.

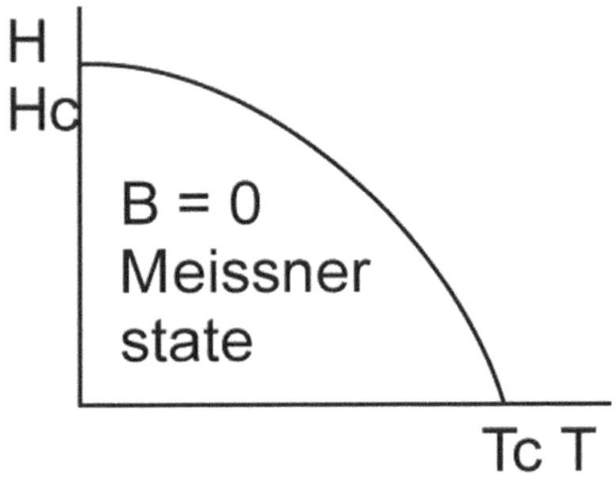

When the field exceeds a lower critical field Hc_1, the magnetization does not drop to zero abruptly rather magnetic flux penetrates the sample partially. When the applied field reaches Hc_2, upper critical field, complete penetration of magnetic flux takes place.

5.2.2 *SPECIFIC HEAT*

At lower temperature, the variation of specific heat of a metal is given by $AT + BT^3$. The first term contributes due to conduction electrons and the second term is due to lattice vibrations. As the temperature of metal is lowered, a discontinuity in specific heat appears at $T = T_c$. The specific heat in the superconducting state is higher than the expected value for the metal. On further lowering the temperature, the decrease of specific heat follows an exponential law given as

$$C_v \propto \exp^{\left(\frac{-\Delta}{K_B T}\right)}$$

It clearly shows that the excited energy level and ground energy level are separated by energy 2Δ in the spectrum of a superconductor. Variation of specific heat of a normal metal and a superconductor with temperature can be best illustrated by the graph.

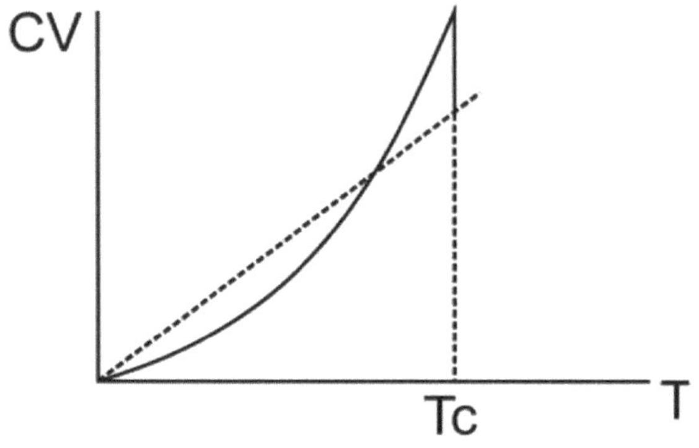

5.2.3 ISOTOPE EFFECT

Formation of cooper pairs takes place by the phonon-mediated attractive electron–electron interaction. The relation between the transition temperature and the ionic mass is given below:

$$T_c \propto \frac{1}{\sqrt{M}}$$

where T_c is critical temperature and M is ionic mass.

Benorze and Muller were awarded Nobel Prize for Physics in 1987 for some of high temperature superconductors (HTSC).

For example, $La_{(2-x)}Ba_xCuO_{4-y}$

$$x = 0.15 - 0.20$$

$$y = \text{small}$$

has a perovskite structure based on La_2CuO_4. $La_2CuO_4^-$ is non-conducting. If 7.5 – 10% of La^{3+} is replaced by Ba^{2+}, La_2CuO_4 becomes superconductor. Similar semiconductors may be prepared by replacing Ba^{2+} with Ca^{2+} or Sr^{2+} and substituting different lanthanides.

Y – Ba – Cu –O SYSTEM

C. W. Chu and M. K. Wu (1987) reported $YBa_2Cu_3O_7$ known as 1–2–3 compound or YBCO to be a superconductor with the critical temperature

93K. The basic structure of the perovskite structure of barium titanate class of ferroelectrics has cations of Ba^{2+} at the corners of the cube, the anions O^{2-} at the center of the faces of the cube. In YBCO, copper ions are at the four corners of the cube while Y and Ba ions alternate at the centers of the cube. Due to tripling of the cube, there are nine possible oxygen sites in the modified perovskite.

Three cubic perovskite units staked one on the top of other, for example,

Cu BaO_3
$CuYO_3$
$CuBaO_3$

For an ideal perovskite structure six metals require Y $Ba_2Cu_3O_9$ composition. Thus 1–2–3 systems are massive oxygen deficient, that is, $\frac{1}{4}$ of the oxygen sites in the crystal are vacant.

5.3 SUPERCONDUCTOR WITH BI AND TL

If Bi and Tl are used instead of lanthanides, superconductors are obtained as

$Bi_2\,Sr_2Ca_{(n-1)}Cu_{\,n}O_{(2n+4)}$ system

N	T_c
1	12K
2	80K
3	110K
4	90K

$Tl_2Ba_2Ca_{(n-1)}Cu_nO_{(2n+4)}$ System

N	T_c
1	90K
2	110K
3	122K
4	119K

5.4 SUPERCONDUCTORS WITHOUT CU OR LANTHANIDE

$BaBiO_3$ has a perovskite structure. It does not behave as a superconductor. If some Ba sites are replaced by K, it behaves as a superconductor, for example, $K_xBe_{(1-x)}BiO_3$.

If some of the Bi sites are replaced by Pb, it behaves like superconductor having a low value of T_c.

Other examples are

$Hg\ Ba_2\ Ca_2\ Cu_3O_8$	T_c (135K)
$Bi_{1.7}\ Pb_{0.2}\ Sb_{0.1}\ Sr_2\ Ca_2\ Cu_{2.8}\ Oy$	(164K)
$Tl_2\ Ba_2\ Ca_2\ Cu_3\ O_{10}$	(T_c 125K)

5.5 THEORY OF SUPERCONDUCTIVITY

Bardeen, Cooper, and Schriefeer suggested the formation of pair of electrons (COOPER pairs). Due to lattice vibrations cooper pairs are formed. When attraction between electrons due to succession of phonon interactions is higher than mutual repulsion of conduction electrons, this theory is known as BCS theory.

5.5.1 LIMITATIONS

The consequence of the BCS theory is that T_c should have an upper limit of 35K. High-temperature superconductor needs modification in the BCS theory or new explanation.

5.5.2 CHARACTERISTICS OF BCS THEORY

i) The critical field, the thermal properties, and most of the electromagnetic properties are results of the energy gap between the ground and excited states. This energy gap is generated due to an attractive interaction between electrons.

ii) When one electron interacts with the lattice and deforms it, a second electron adjusts itself to take advantage of the deformation to lower its energy. In this way the second electron interacts with the first election via lattice deformation.

iii) The penetration depth and the coherence length emerge as a result of the BCS theory. Thus, the Meissner effect is obtained in a natural way.

iv) The criterion for transition temperature is *UD* (ε*F*) << *I* as per the BCS theory.

$$T_c = 1.14\theta \exp^{\left(\frac{-1}{UD(\varepsilon F)}\right)} \; \mathrm{T_C} = 1.14\theta$$

where θ is the Debye temperature, *U* is an attractive interaction, and *D*(ε*F*) is the electron density of orbitals of one spin at the Fermi level and the electron lattice interaction *U*.

The higher the resistivity at room temperature, the higher is *U*. Thus, it is probable that metal will be a superconductor when cooled.

v) The BCS ground state involves pairs of electrons. Magnetic flux through a superconducting ring is quantized and the effective unit of charge is $2e^-$ instead of e^-.

5.6 ORGANIC SUPERCONDUCTORS

Generally organic solids are insulators. But a number of organic solids now exhibit conductivity like metals. TTF–TCNQ type material shows unusual conductivity of 5000 $\Omega^{-1}cm^{-1}$ at 85K. Superconductivity has been found in a class of selenium-based organic compounds at a critical temperature of around 1.5K. Charge transfer salt of $(TMTSF)_2X$ also shows such properties.

TMTSF stands for tetramethyl tetra Selena fulvalene and X is an inorganic anion such as PF_6, TaF_6, ReO_6 etc.

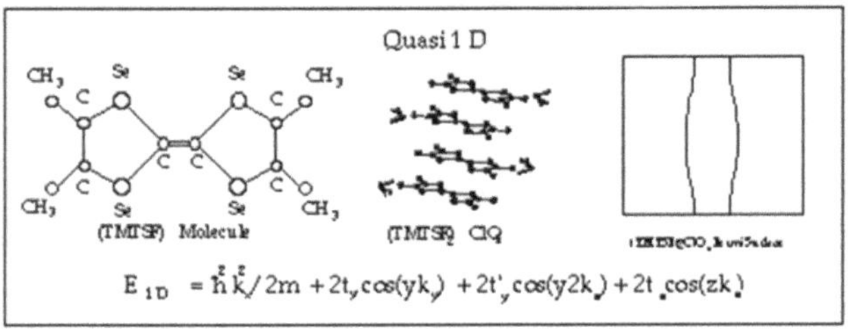

Tetramethyl tetra selena fulvalene (TMTSF)

Several members of this family show superconducting behavior.

5.6.1 FULLERENES

The archetype fullerene C_{60} is a cage-like molecule. Each molecule has the form of truncated icosahedrons with 20 hexagonal faces and 12 pentagonal faces, like a soccer ball. All the angles are equal and their faces are regular polygons.

C_{60} is itself not a superconductor. When alkali metals are added to it, it can become superconducting. The C_{60} molecule is packed in an fcc structure with a lattice constant 14.2 Å. Since the radius of the molecule is 3.54 Å, the vacancy is enough to accommodate dopants of small radii. Thus alkali-fullerene compounds such as K_3C_{60} are superconducting at T_c of 19.2K. The potassium atoms occupy the octahedral sites in the cubic cell. Rb_3C_{60} have also been reported to show superconductivity with a critical temperature of about 30K. $RbCs_2C_{60}$ is also a superconductor having T_c 33K.

5.6.2 APPLICATION

i) Power transmission without any loss in transmission can be realized.

ii) Due to zero resistance, heat problems are reduced.

iii) **Powerful electromagnet:** Normal conductor bears current density of $10^7 A/m^2$ order and superconductor bears $10^{10} A/m^2$. High current can be passed through a very fine wire of superconductor with a large number of turns to produce extremely powerful electromagnet.

iv) Powerful magnets can be used in linear accelerators used as atom smashers in high-energy particle research physics.

v) Superconducting magnets of 2 Tesla are used in MRI.

vi) On the basis of the property of levitation, an experimental frictionless train has been developed. This train floats above 500 km/h.

KEYWORDS

- **Meissner effect**
- **BCS theory**
- **organic superconductors**
- **isotope effect**
- **specific heat**
- **fullerenes**

CHAPTER 6

Thermal Properties of Solids

ABSTRACT

The lattice heat capacity of a crystal has been explained by Einstein's model and later on modified by Debye model. Debye T^3 law satisfies the experimental results at low temperatures. Vibrations of crystals with monoatomic basis have been considered. Elastic response of the crystal is a linear function of the forces. Brillouin zone, group velocity and phonon momentum have also been discussed in this chapter.

6.1 INTRODUCTION

Heat capacity is also the thermal property of metals. Though heat capacity at constant volume and heat capacity at constant pressure are used, heat capacity at constant volume is more fundamental than heat capacity at constant pressure. Heat capacity at constant volume is defined as $C_v = \left(\frac{\delta E}{\delta T}\right)_v$, where E is the energy and T is the temperature. Heat capacity of a crystal is generated due to the contribution of the phonons and so-called lattice heat capacity and is expressed as C-lattice.

At certain temperature T, the total energy of the phonons in a crystal may be shown as the sum of the energies over all phonon modes. If K is a wave vector and p is a polarization index, $E = \sum_k \cdot \sum_p \cdot \sum_{k}, p$

6.2 EINSTEIN MODEL

i) The atoms in a crystal lattice undergo small oscillations about their equilibrium configurations.

Solid-State Chemistry: A Modern Approach. Ashok Kumar Jha, PhD (Auth.)

ii) Each atom vibrates independently and has three independent vibrational degrees of freedom. Let the crystal be a system of 3N distinguishable harmonic oscillators.

Einstein thus considered a solid as an aggregate of atomic oscillators. Each of these oscillators is vibrating with a common mean frequency v.

On the basis of the quantum theory, the energy of an oscillator, vibrating with v in one direction, is given by $\varepsilon = \frac{hv}{e\frac{hv}{kT} - 1}$ where h is Planck's constant, k is Boltzmann constant.

Let the degree of freedom be 3 and No total number of atoms in a gm atom

Total energy of a gm atom of solid is

$$E = 3No.\frac{hv}{e\frac{hv}{kT} - 1}$$

Heat capacity

$$C_v = \left(\frac{\delta E}{\delta T}\right)_v = 3Nok\left(\frac{hv}{kT}\right)^2 \left(e\frac{hv}{kT} - 1\right)^2$$

$$= 3R\left(\frac{hv}{kT}\right)^2 \left(e\frac{hv}{kT} - 1\right)^2$$

Putting $\frac{hv}{k} = \theta$, we have

$$C_v = 3R\left(\frac{\theta}{T}\right)^2 \frac{e^{\frac{\theta}{T}}}{\left(e^{\frac{\theta}{T}} - 1\right)^2}$$

Thus C_v is a function of $\left(\frac{\theta}{T}\right)$ and may be written as

$C_v = 3R\int\left(\frac{\theta}{T}\right)$, where θ depends upon v and is a characteristics of the solid when temperature is high, $\frac{hv}{k_B T}$, is quite small $C_v = 3R$

This is in agreement with the classical theory.

In a lower range of temperature, the calculated values fall more swiftly than the experimental values. This disagreement may be due to interatomic forces.

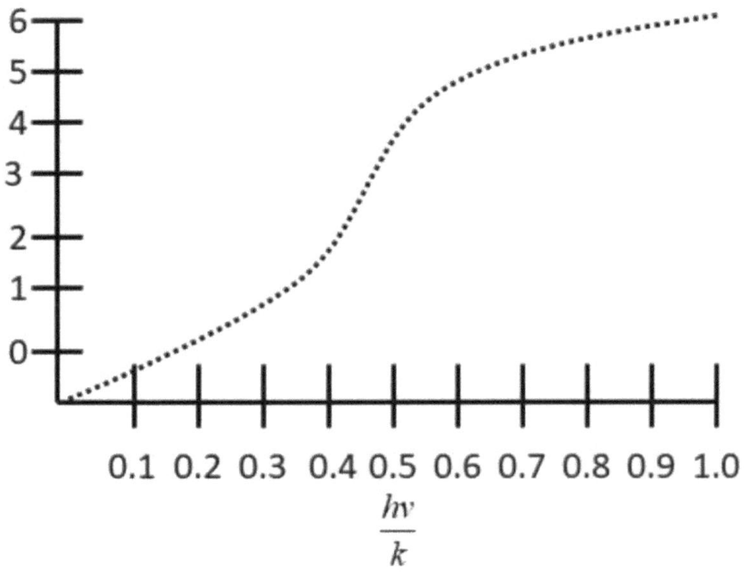

FIGURE 6.1 Heat capacity versus characteristic temperature.

Comparison of experimental values of the heat capacity of diamond with values calculated on the earliest Einstein model using the characteristic temperature $\theta = \frac{hv}{k} = 1320K$

Thus, the Einstein model is applied to explain the optical phonon part of the spectrum of phonons.

6.3 DEBYE MODEL FOR DENSITY OF STATES

According to Debye, a solid is assumed to be an elastic body and the vibration of the whole should be considered. Modes of vibration of a monoatomic solid are not all identical but are distributed among a spectrum of frequencies. The frequency ranges from $(v = o)$ to $v - V_m$ (maximum frequency).

Debye's V_m is different from Einstein's V

Total number of ways is z in which an elastic medium of volume V can vibrate with a frequency v, and then dz between the ranges $v + dv$ and v will be given by

i) $dz = \frac{4\pi V}{c^3}.v^2 dv$ (6.1)

Let No be the atoms in a gm-atom of the solid element.

The total number of vibrations possible will be 3 No.

$$\int dz = \int_0^{V_m} \frac{4\pi V}{c^3}.v^2 dv \tag{6.2}$$

$$= \frac{4\pi vm^3}{3c^3}.V = 3No \tag{6.3}$$

From (6.1) and (6.3) eliminating V, we have

$$dz = \frac{9No}{Vm^3}.v^2 dv$$

Energy associated with each linear oscillator is given by

$$\varepsilon = \frac{hv}{e^{\frac{hv}{KT}-1}} \tag{6.4}$$

The total energy of a gm-atom of solid will be

$$E = \int_0^{V_m} \varepsilon dz$$

$$= \int_0^{V_m} \frac{hv}{e^{\frac{hv}{KT}-1}}.\frac{9No}{Vm^3}.v^2 dv$$

$$E = \frac{9Noh}{Vm^3}\int_0^{V_m} \frac{v^3}{e^{\frac{hv}{KT}-1}}.dv \tag{6.5}$$

Here Debye's equation adheres to classical. Let $x = \frac{hv}{KT}, v = \frac{KT}{h}x, v^3$ $dv = \left(\frac{KT}{h}\right)^4 x^3 dx$

From eq (6.5)

$$E = \frac{9No(KT)^4}{(hV_m)^3}\int_0^{x_m} \frac{x^3 dx}{e^x - 1} \tag{6.6}$$

where $x_m = \frac{hv_m}{KT}$

a) When the temperature is high, x is small $\int_0^{x_m} \frac{x^3 dx}{e^x - 1} = \frac{1}{3} x_m^{\ 3}$

Putting into eq (6.6) we have

$$E = \frac{9No(KT)^4}{(hV_m)^3} \frac{1}{3} . x_m^{\ 3}$$

$$= 3\ NoKT = 3RT$$

$$C_v = \left(\frac{\delta E}{\delta T}\right)_v = 3R$$

treatment and agrees with experimental values at higher temperature.

b) At low temperature

$$\int_0^{x} \frac{x^3 dx}{e^x - 1} = \frac{\pi^4}{15}$$

$$E = \frac{9No(KT)^4}{(hV_m)^3} \frac{\pi^4}{15}$$

$$= \frac{3No}{5} \frac{(\pi KT)^4}{(hV_m)^3}$$

$$C_v = \left(\frac{\delta E}{\delta T}\right)_v = \frac{12}{5} \pi^4 No.K \left(\frac{KT}{hV_m}\right)^3$$

$$= \frac{12}{5} \pi^4 R \left(\frac{T}{\theta}\right)^3$$

$$C_v = 464.6 \left(\frac{T}{\theta}\right)^3$$

where $\frac{hV_m}{K} = \theta$ (characteristic temperature of the substance).

$$C_v = \frac{464.6}{\theta^3} . T^3$$

$$\text{or } C_v\ \alpha\ T_3$$

This equation is known as Debye's T^3-law.

At a sufficiently low temperature the T^3 approximation is quite good.

i) Heat capacity in $J.mol^{-1}K^{-1}$

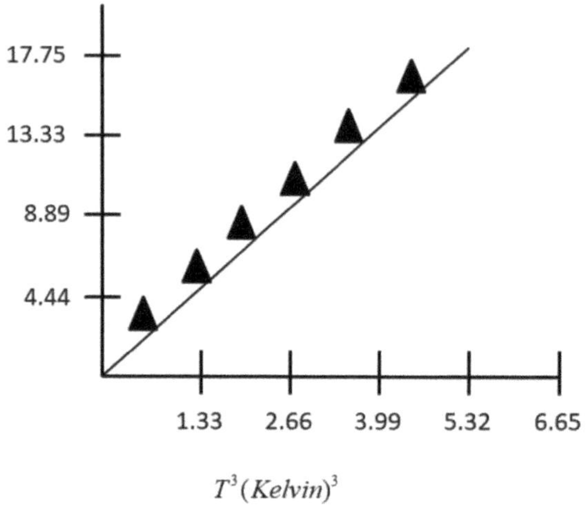

FIGURE 6.2 Heat capacity versus T^3.

When heat capacity of solid argon at low temperature is plotted against temperature region, the experimental values agree with Debye T^3 law with $\theta = 92.0\ K$

Heat capacity in $J.mol^{-1}.K^{-1}$

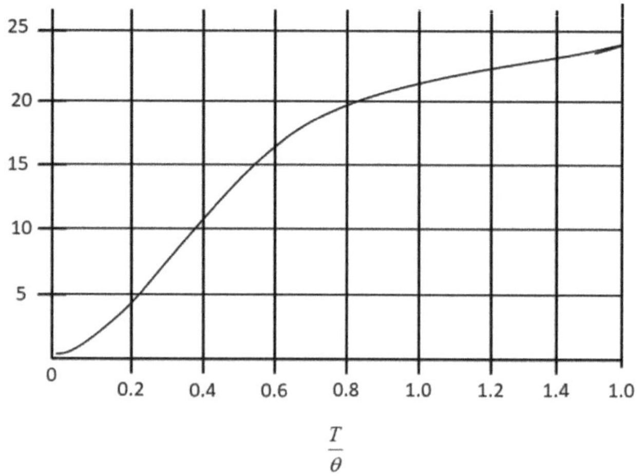

FIGURE 6.3 Heat capacity versus $\frac{T}{\theta}$.

Figure 6.3 shows heat capacity C_v of a solid according to Debye approximation. Heat capacity on Y-axis is in $J.mol^{-1}k^{-1}$. The X-axis represents the temperature normalized to the Debye temperature θ. The region of the T^3 law is below 0.1θ. The asymptotic value at a high value of $\frac{T}{\theta}$ is 24.943 $J.mol^{-1}deg^{-1}$

Thus, the Debye equation predicts that the heat capacity of a solid is a function of temperature and depends only on the characteristic maximum frequency V_m.

6.4 PHONONS

Vibration of the atoms takes place to and fro their equilibrium position. Lattice vibrations are represented by waves. The energy of lattice vibrational waves is quantized. This quantum is called a phonon. Transmission of a displacement wave in a solid may be said as the movement of one or more phonons each of which has energy (hv) and momentum. Thermally excited phonons are nothing but thermal vibrations.

6.4.1 *PHONON MOMENTUM*

The phonons have no momentum in itself. But a phonon of wave vector K interacts with photons, neutrons, and electrons.

Phonons in a lattice have no momentum because of involvement of relative coordinates of the atoms with phonon coordinate.

For example, in an H_2 molecule the inter-nuclear vibrational coordinate r_1-r_2 is a relative coordinate and hence has no linear momentum.

6.4.2 *GROUP VELOCITY*

Group velocity may be defined as the transmission velocity of a wave packet.

$$V_g = \frac{d\omega}{dK}$$

or $V_g = \nabla k\omega(K)$

the frequency gradient with respect to K.

From eq (6.15)

$$V_g = \left(\frac{Ca^2}{M}\right)^{\frac{1}{2}} \cos\frac{1}{2}ka$$

This is zero at the edge of the zone where $k = \frac{\pi}{a}$

For a standing wave the net transmission velocity is expected to be zero.

6.5 ONE-DIMENSIONAL MONOATOMIC LATTICE

Elastic vibrations in the primitive cell having one atom have been considered here. If the wave propagates along either of the directions [100], [110], or [111], whole planes of atoms move in phase with displacements either parallel or perpendicular to the wave vector. The displacement of the planes from its equilibrium position can be explained with a single coordinate Us. There are three modes for each wave vector, one longitudinal polarization and two transverse polarizations.

Figure	**Name**	**Field**
→	Electron	
	Photon	Electromagnetic wave
	Phonon	Elastic wave
	Magnon	Magnetization wave
———	Exciton	Polarization wave

FIGURE 6.4 Important elementary excitations in solids.

The sum of total forces on S from planes S ± 1

$$Fs = C\,(Us{+}1 - Us) + C\,(Us{-}1{-}Us)$$

This expression is linear in displacements and is on the pattern of Hooke's law.

C is the force constant between the neighbor planes for the longitudinal and transverse waves.

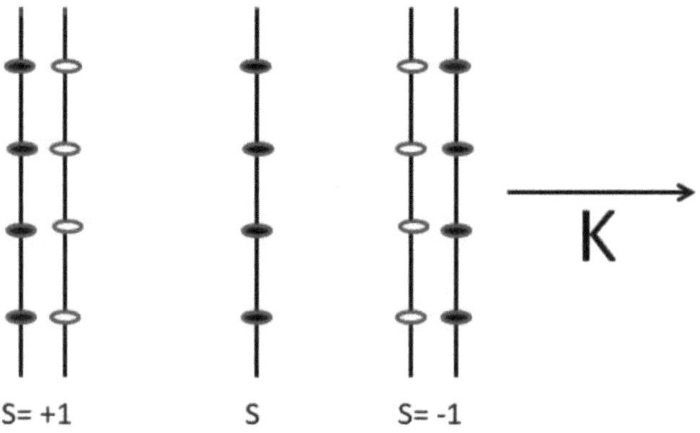

FIGURE 6.5 Planes of atoms at equilibrium represented by solid lines, planes of atoms after displacement for a longitudinal wave.

The equation of motion of the planes is

$$M.\frac{d^2us}{dt^2} = c(us+1+us-1-2us) \quad (6.7)$$

where *M* represents the mass of an atom

If all displacements have the time dependence exp (-iwt), then

$$\frac{d^2us}{dt^2} = -w^2us \quad (6.8)$$

Putting into eq (6.7) we have

$$-MW^2U_s = c(us+1+us-1-2us) \quad (6.9)$$

This is a differential equation in the displacement U and the solution is of the form

$$us \pm 1 = u\exp(ika)\exp(\pm\, ika) \quad (6.10)$$

where *a* shows the spacing between planes and K shows the wave vector.

From (6.9) and (6.10)

$$-w^2Mu\exp(iska) = cu\{\exp[i(s+1)]ka\} + \exp[i(s-1)Ka] - 2\exp(iska) \quad (6.11)$$

$$w^2M = -c[\exp(-ika) + \exp(ika) - 2] \quad (6.12)$$

From

$$\exp^{ika} + \exp^{-ika} = 2\cos ka$$

Eq (6.12) now becomes

$$w^2 = \left(\frac{2c}{M}\right)(1 - \cos ka) \tag{6.13}$$

From eq (6.13) the slope of w versus k is zero at the zone boundary.

$$\frac{dw^2}{dk} = \frac{2ca}{M}\sin ka = 0 \tag{6.14}$$

The extreme boundary in the first Brillouin zone is at $k = \pm\frac{\pi}{a}$
Eq (6.13) can be written as

$$w^2 = \frac{4c}{M}\sin^2\frac{1}{2}ka$$

or,

$$w = \left(\frac{4c}{M}\right)^{\frac{1}{2}} \sin\frac{1}{2}ka \tag{6.15}$$

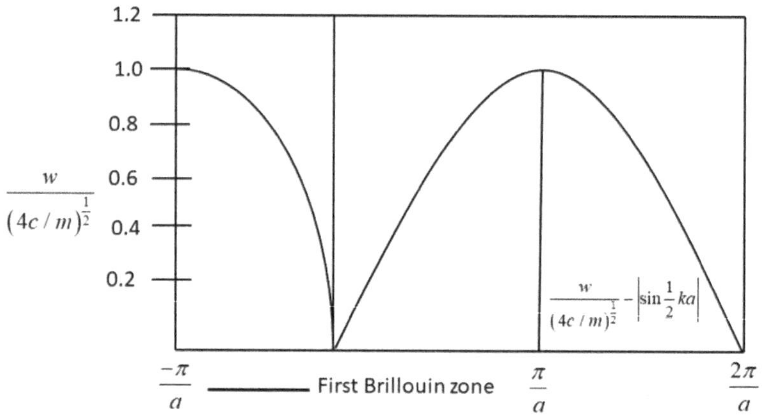

FIGURE 6.6 Plot of w versus k.

The displacements ratio of two sequential planes may be shown by

$$\frac{u_s + 1}{U_s} = \frac{u\exp[i(s+1)ka]}{u\exp(iska)} = \exp(ika) \tag{6.16}$$

The phase range $-\pi$ or π for the phase K_a envelops all self-supporting values of the exponential.

Thus the range $-\pi$ or π for the phase Ka covers all self-supporting values of the exponential. Thus, the range of self-supporting values of K is specified by

$$-\pi < K_a \leq \pi$$

$$or, -\frac{\pi}{a} < K \leq \frac{\pi}{a}$$

This array is the first Brillouin Zone in the linear lattice

The extreme values are $K_{\max} = \pm\frac{\pi}{a}$

6.6 QUANTIZATION OF ELASTIC WAVE

The energy of a lattice vibration, called phonon, is quantized. Elastic waves made up of phonons propagate in crystals. In black-body electromagnetic radiation there are thermally excited photons.

Let v be the frequency of the phonons and E be the energy.

$$E = \left(n + \frac{1}{2}\right)hv$$

where n is quantum number

If $n = 0, \frac{1}{2}$ represents the zero point energy of the mode.

Thus for quantum harmonic oscillator of frequency v, the energy eigenvalues are $\left(n + \frac{1}{2}\right)hv$, equally spaced in a harmonic oscillator.

The energy in the mode in any harmonic oscillator is supposed to be contributed by kinetic energy and potential energy equally.

6.7 ANHARMONIC CRYSTAL INTERACTIONS

Harmonic theory of crystal interactions has the consequences given below.

i) A single wave does not change form with time.
ii) Thermal expansion does not take place.
iii) At high temperature the heat capacity remains constant.
iv) The elastic constants do not vary with pressure and temperature.

But real crystals behave like a harmonic oscillator. So the consequences are not followed.

6.8 THERMAL CONDUCTIVITY

The thermal conductivity of a solid is defined as flow of heat down a rod with a temperature gradient $\frac{dT}{dx}$.

$$J_E = -K\frac{dT}{dx} \tag{6.17}$$

where J_E stands for the flux of thermal energy.

Flux may be defined as the energy transmitted per unit area per unit time. The energy diffuses as a result of frequent collisions which implies that transfer of thermal energy is random. Due to randomness in the conductivity process, temperature gradient develops.

Debye explained thermal conductivity in dielectric solids by the equation

$$K = \frac{1}{3}CVl \tag{6.18}$$

where K represents the thermal conductivity, V represents the average particle velocity, C is the heat capacity per unit volume, and l is the mean free path.

Putting the value of K from eqs (6.18) in (6.17), we have

$$J_E = -\frac{1}{3}CVl.\frac{dT}{dx}$$

Dielectric crystals have also high thermal conductivity like metals. Al_2O_3 has been found to possess high conductivity.

KEYWORDS

- **Brillouin zone**
- **Einstein's model**
- **Debye T^3 law**
- **group velocity**
- **phonon momentum**
- **monoatomic basis**

CHAPTER 7

Energy Bands: Metals, Insulators, Semiconductors, and Organic Solid

ABSTRACT

Metals, semiconductors and insulators are classified by their electrical resistivity. The band gap indicates the energy difference between conduction band and valence band. Band theory of solids explains the behavior of metals, semiconductors and insulators. Extrinsic semiconductors of n and p type have been explained. The temperature dependence for a semiconducting material has been shown by figure. Interpretation of semiconductors has also been given by free electron theory. Brillouin zone has been explained and a mathematical interpretation of band theory has been dealt with.

7.1 INTRODUCTION

Though the free electron model of metals explains heat capacity, thermal conductivity, and electrical conductivity, it fails to explain the difference between, metals, insulators, and semiconductors. The main difference between metals, semiconductors, and insulators is the difference in their conductivities .The conductivities are given below

Metal—10^4 to 10^6 $ohm^{-1}cm^{-1}$
Insulators—10^{-15} $ohm^{-1}cm^{-1}$
Semiconductors—10^{-5} to 10^3 ohm^{-1} cm^{-1}

There is an increase in the conductivity of semiconductors, but there is a decrease in the conductivity of metals with rise in temperature.

Solid-State Chemistry: A Modern Approach. Ashok Kumar Jha, PhD (Auth.)

Let σ be the conductivity
n be the number of electrons
e be the charge
μ be the mobility of charge carriers.
$\sigma = ne\mu$

In case of metals n is large and remains unchanged with rising temperature, but μ the mobility decreases slightly with increasing temperature due to collisions between the moving electrons and phonons.

The main reason for different behavior in metal, insulator, and semiconductor is n.

In case of metals n is large and remains unchanged with rising temperature, but μ the mobility decreases with temperature. In case of insulators and semiconductors n increases with temperature and hence conductance σ rises with increasing temperature. This is why some insulators become semiconductor at high temperature and semiconductors become insulator at low temperature.

Semiconductors can be classified as follows.

7.2 ELEMENTAL SEMICONDUCTOR

Silicon, germanium, and gray Sn are the classic semiconductors.

All have the diamond structure in which each atom is surrounded tetrahedrally by four other atoms. The tetrahedral link joins together to form a rigid network.

7.3 COMPOUND SEMICONDUCTORS

Many inorganic compounds of Groups III–V have a combination of 1:1, for example, gap shows semiconductor behavior. These types of compounds have the zinc blende structure. Many organic compounds, for example, polyacetylene, TTF—TCNQ show semiconductor behavior.

7.4 BAND THEORY AND MOLECULAR ORBITAL THEORY

Molecular orbital theory explains band theory. In the molecular orbital theory of molecules, for example, atomic orbital from atom 1 overlaps with atomic orbital from atom 2 which results in the formation of bonding and antibonding molecular orbital. The energy of bonding molecular orbital is lower than the energy of atomic orbitals and the energy of antibonding molecular orbital is higher.

In case of larger molecules there is an increase in the number of molecular orbitals. When the number of molecular orbitals increases, the gap of energy between adjacent molecular orbitals decreases.

A situation is reached in which there is a continuum of energy

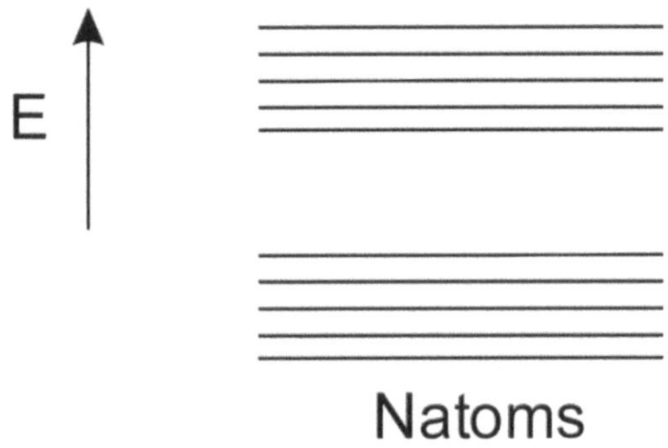

The band structure of metallic sodium is given below.

At the experimental value for interatomic separation r_o, the 3s and 3p orbitals overlap to form broad 3s and 3p bands. The upper level of 3s band has similar energies to the lower levels of the 3p band.

At the interatomic distance or, the 1s, 2s, and 2p orbitals on adjacent sodium atoms do not overlap. If sodium is compressed, the internuclear separation r_o may decrease to r'. In such a case the 2s and 2p orbitals would overlap to form bands of energy levels. These energy levels would remain intact. At high pressure similar effects can be observed. It has been calculated that hydrogen would become metallic at a pressure of 10^6 atm.

The valence electron of sodium is not confined to the 3s band, but it is distributed over the lower levels of both the 3s and 3p bands. Occurrence of K β line in the X-ray emission spectra of sodium metal can be explained.

7.5 FREE ELECTRON THEORY

A metal is regarded as a potential well having free valence electron for movement. The energy levels are quantized which are filled from the bottom of the well with two electrons per level. The Fermi level is the highest filled level at absolute zero.

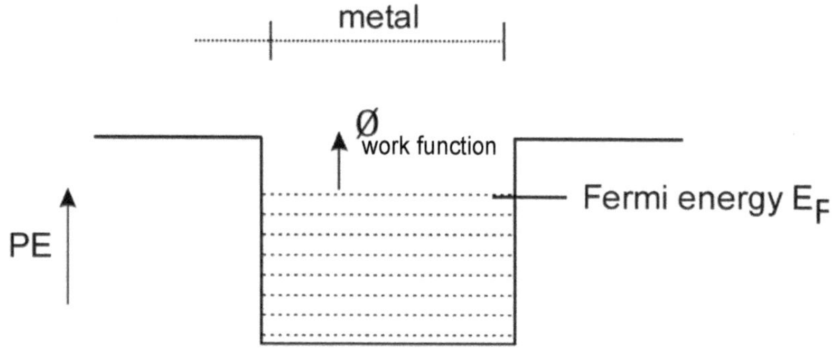

The work function Ø is the energy required to remove the valence electron from the potential well.

In a more refined theory, the potential inside the crystal is not constant but periodic. The potential energy of the electrons passes through a maximum midday between adjacent nuclei.

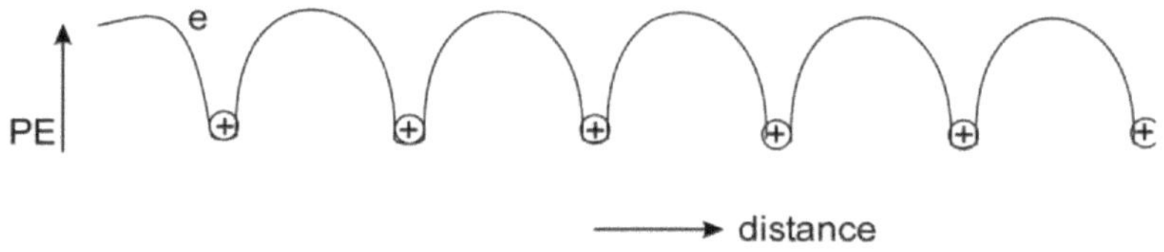

Potential energy of electrons vs distance through a solid

Thus, the existence of energy bands in solids is obtained from both molecular orbital and periodic potential approach. There are bands of energy levels for the valence electrons. In some materials overlap of different bands occurs and in some, a forbidden gap exists between energy bands.

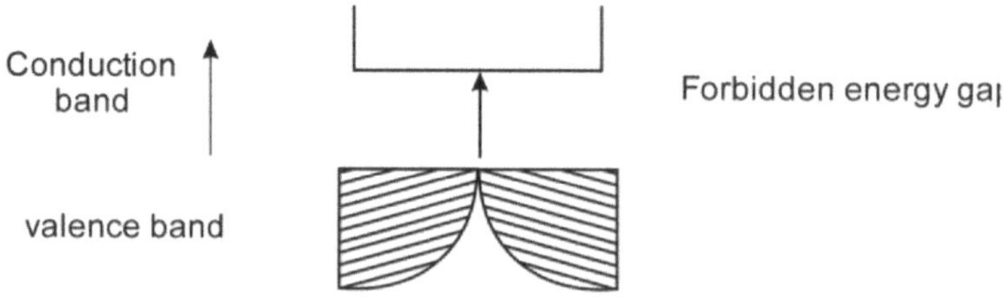

k space and Brillouin zones

7.6 SIMPLE BAND THEORY

It has been assumed that periodic variation of energy takes place on the movement of electron through the crystal. The expression for the value of Ψ was obtained by Bloch.

The Schrödinger equation when a free electron moves in the x-direction is

$$\frac{d^2\Psi}{dx^2}+\frac{8\pi^2 m}{h^2}(\mathrm{E}-\mathrm{V})= 0$$

This leads to the conclusion

$$\Psi = \mathrm{Ce} \pm 2\,\pi\,\mathrm{i}\frac{\sqrt{2m\,(E-V)}}{h}.x$$

The kinetic energy of a particle such as a free electron is given by

$$\mathrm{E} = \frac{1}{2}\mathrm{mv}^2 \tag{7.1}$$

From wave particle duality

$$\lambda = \frac{h}{mv} \tag{7.2}$$

vector k whose direction is parallel to the direction of propagation of the waves is related to reciprocal wavelength by

$$K = \frac{2\pi}{\lambda} \tag{7.3}$$

$$\text{or } \lambda = \frac{2\pi}{K}$$

$$\text{or, } \lambda = \frac{2\pi}{K}$$

$$\mathrm{E{-}V} = \frac{1}{2}\mathrm{mv}^2$$

$$= \frac{1}{2}\frac{(mv^2)}{m}$$

$$= \frac{1}{2m}(\frac{h}{\lambda})^2 = \frac{1}{2m}\cdot\frac{h^2}{\lambda^2}$$

$$= \frac{1}{2m} \cdot \frac{h^2}{\left(\frac{2\pi}{k}\right)^2}$$

$$\text{E-V} = \frac{h^2 K^2}{8\pi^2 m} \quad (7.4)$$

$$= \frac{h^2 K^2}{8\pi^2 m}$$

Substituting this into Schrödinger's equation, the general solution for a free equation making at constant potential may be obtained:

$$\Psi(x) = Ce^{\pm ikx}$$

According to the band theory potential energy V does not remain constant rather it varies with Z

$$\frac{d^2\Psi}{dx^2} + \frac{8\pi^2 m}{h^2}(\text{E} - \text{v}(x))\Psi$$

Bloch has put forward the solution of this equation

$$\Psi(x) = \text{u}_\text{k}(x)\text{e}^{\pm \text{ikx}}$$

where u_k is a periodic function. When the electron moves along the x direction, the magnitude of (U_K (x) varies, but the value remains unaltered at two points segregated by the lattice spacing.

$$\text{u}_\text{k}(x + \text{a}) = \text{u}_\text{k}(x)$$

where a is the lattice constant.

Kroning and Penney assumed that the potential energy is zero near the nucleons and maximum Vo at the point half way between two neighboring nuclei. They assumed that the product of the potential height and width ($V_o w$) is constant.

The obtained relation $\cos k\alpha = \frac{P \sin \alpha a}{\alpha a} + \cos \alpha a$

$$\text{When P} = \frac{4\pi^2 m\alpha}{h^2} V_o.W$$

$$\alpha = \frac{2\pi}{h}\sqrt{2mE}$$

Now $\cos k\,\alpha$ can vary only between +1 and–1 if k is real, implying that all energy values are not permissible.

The shaded regions show the allowed values of cos kα. The allowed values of the electron energy can be calculated from the relation between α and E

When the allowed zones P → α, switch to discrete levels. This happens as α a = nπ where n is a natural number. When P → O, the forbidden regions disappear and the electron is like a free particle.

The intervals of k within which E completes its full cycle are known as Brillouin zone. The first Brillouin zone lies between K = $\frac{-\pi}{a}$ and K = $\frac{\pi}{a}$ and its width is $\frac{2\pi}{a}$

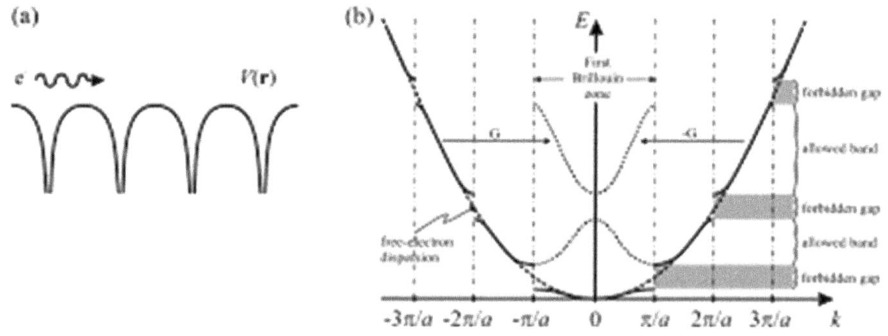

FIGURE Brillouin zone

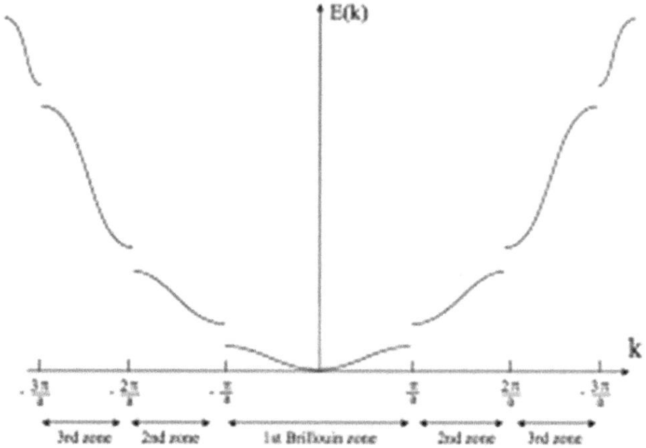

FIGURE Brillouin zone

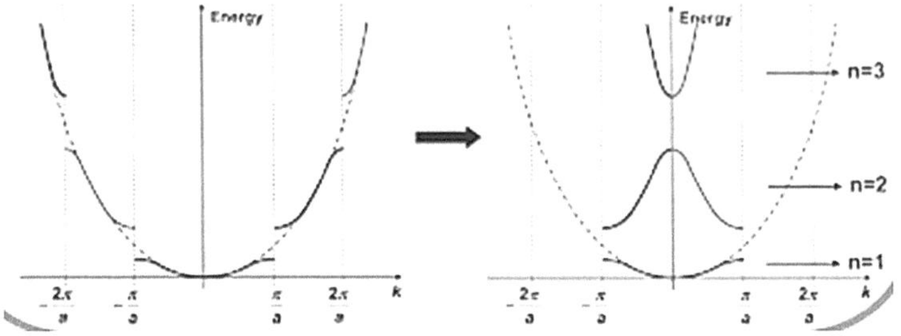

View of Brillouin zone

Each of the planes (100), ($\bar{1}$00), (010), (0$\bar{1}$0), (001), and (00$\bar{1}$) contributes a face to the 100 Brillouin zone which takes the forms of a cube in three dimensions. The surface of the Brillouin zone is parallel to the diffracting planes (100).

Thus, the first Brillouin zone in a primitive cubic structure corresponds to diffraction from the (100) planes. The second Brillouin zone is diffraction from the (110) planes .There are twelve sets of planes in the set (110). The Brillouin zone for this takes the shape of a regular dodecahedron that envelops the cube shaped.

For bcc, the first Brillouin zone has the same dodecahedron corresponding to reflection from the (110) planes.

For a face-centered cubic, the first Brillouin zone takes the shape of a truncated octahedran with fourteen faces.

Now the energy of valence electrons and their positions in k space are considered. In metals or semiconductors electrons lie deep down in the valence band. A metal or semiconductor has low energy. Its k value is also very low. In k space, they would be inside the first Brillouin zone. As the energy of the electrons increases, the k vector occupies the first Brillouin zone when k brings to a close on the zone boundary, a forbidden condition is obtained.

If the energy of electrons increases further, the k vectors pass through the boundary and enter the second zone.

7.7 BAND ENERGY OF METALS

For metals, the highest occupied band, that is, the valence band is filled partly. Some levels just below the Fermi level are vacant whereas some above are occupied.

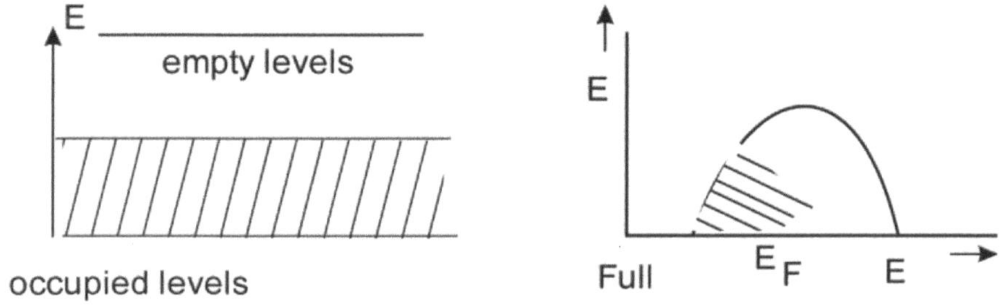

FIGURES 7.1 Band structure of a metal.

Electrons in singly occupied states close to E_F are responsible for the high conductivity of metals due to movements of electrons. In the band structure of Be, it has overlapping 2s, 2p bands, both are only partly full. If there is no overlapping between the 2s and 2p bands in Be, the 2s band would be full and the 2p band would be empty. In such a case Beryllium would not be metallic.

7.8 SEMICONDUCTORS

i) Intrinsic semiconductor
ii) Extrinsic semiconductor

7.8.1 INTRINSIC SEMICONDUCTOR

Silicon and Ge of group 14 have narrow band gaps. At room temperature, electrons are excited thermally to cross the band gap between the valence band and the conduction band. Conduction is due to the movement of electrons and holes in the conduction band and valence band respectively. The intrinsic conductivity depends on the ratio of the band gap to the temperature. When this ratio is large, conductivity will be low. At OK the conductivity is zero because all states in the valence band are vacant. As the temperature is increased, electrons move faster from the valence band to conduction bands and so conductivity becomes significant.

The band gaps are obtained accurately by optical absorption. The band gap $E_g = hw_g$ where w_g is the frequency of absorption. In direct absorption, a photon is absorbed by the crystal and electron and a hole are created. In the indirect absorption, the minimum energy gap of the band structure involves electrons and holes separated by a wavefector k_c. The indirect transition involves both a photon and phonon because the band fringes in the conduction and valence bands are separated in the k space. The absorption threshold for the indirect transition between the band edges is at $hw_g = E_g + h\Omega$, where Ω is the frequency of an emitted phonon of wave vecto $r \cong -k_c$

TABLE 7.1 Energy Gap Between the Valence and Conduction Band (i = Indirect Gap, d = Direct Gap).

Crystal	Gap	Eg, ev
Diamond	i	5.4 (0K)–(300K)
Si	i	1.17 (0K)–1.11 (300K)
Ge	i	0.744 (0K)–0. 66 (300K)
InSb	d	0.23 (0K)–0.17 (300K)
InAs	d	0.43 (0K)–0.36 (300K)
InP	d	1.42 (0K)–1.27 (300K)
GaP	i	2.32 (0K)–2.25 (300K)
GaAS	d	1.52 (0K)–1.43 (300K)
GaSb	d	0.81 (OK)–0.68 (300K)
Sn Te	d	0.3 (0K)–0.18 (300K)
CdS	d	2.582 (0K)–2.42 (300K)
HgTe (Semimetal)	d	0.30 (0K)–
PbS	d	0.286 (K)–0.35–0.37 (300K)

7.8.1.1 EXPLANATION OF INTRINSIC SEMICONDUCTOR

The band structure of Si and Ge is quite different to that of Na and Mg. The explanation of the band structure of Si lies in quantum mechanics. The simplest explanation is the assumption that each silicon forms four bonds arranged tetrahedrally, each hybrid orbital overlaps with a similar orbital on an adjacent silicon to form one bonding and one antibonding molecular orbital. Individual molecular orbitals overlap to form bond which is valence band. In this way many bonding molecular orbitals come close and overlap as a result of which valence band is formed. The antibonding molecular orbitals overlap by coming close to each other and as a result conduction band is obtained. The valence band remains filled and conduction band empty. As the temperature is increased, electrons go to conduction band by overcoming to band gap and holes are created in the valence band which gives rise to conductivity

7.8.2 EXTRINSIC SEMICONDUCTORS

7.8.2.1 P-TYPE SEMICONDUCTOR

Extrinsic semiconductors are those in which dopant of Group III or Group V is added to Group IV. For example, silicon is doped with a small amount 10^{-2} atom % of Gallium. In pure silicon Si-Si bond is a pure covalent bond having two electrons. There are four Si-Si bonds or four covalent bonds. Gallium has only three valence electrons. So one of the Ga-Si bonds is deficient by one electron.

Thus, one Ga-Si bond deficient by one electron does not form part of the valence band of Si, but it forms a discrete level or atomic orbital just above the valence band. This level is known as an acceptor level and the gap between the acceptor level and the top of the valence band is only 0.1 eV. It is not possible for electrons in the acceptor levels to contribute to conduction. Electrons from the valence band may have sufficient thermal energy to be promoted to acceptor levels. The positive holes left in the valence band also move and thus gallium/doped silicon is a positive hole or p-type semiconductor.

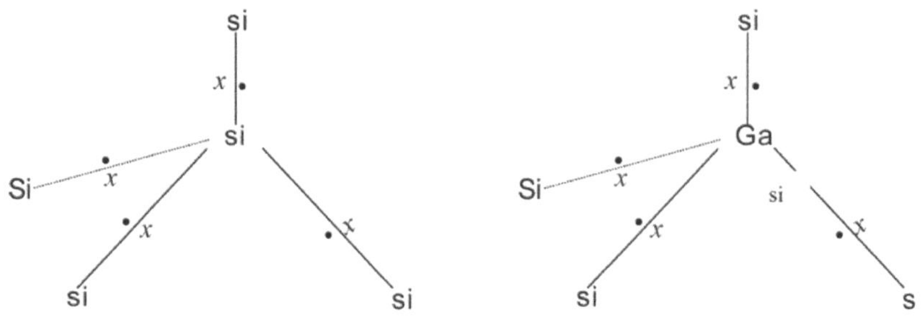

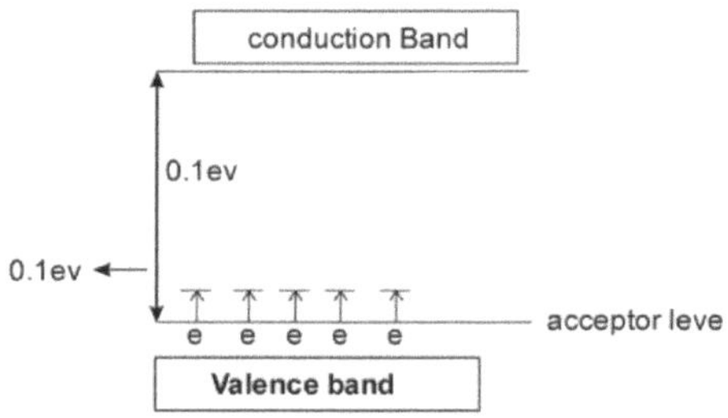

P type semiconductor

The number of positive holes created by Gallium dopant far exceeds the number created by the thermal promotion of electrons into the conduction band. So the conductivity is affected by the concentration of gallium atoms. When the temperature is increased, the concentration of intrinsic carriers increases rapidly. At high temperature, the intrinsic carrier concentration may exceed the extrinsic value.

7.8.2.2 N-TYPE SEMICONDUCTOR

When silicon is doped with a pentavalent element arsenic, there is one electron more than required to form Si-As covalent bond. This extra electron occupies a discrete level that lies below the bottom of the conduction band. This lies 0.1 eV below the rear end of the conduction band. These levels act as donor levels. The electrons in these levels have sufficient thermal energy to go to the conduction band. Due to the electrons in the conduction band, conductivity arises. Materials possessing such properties are called n-type semiconductor.

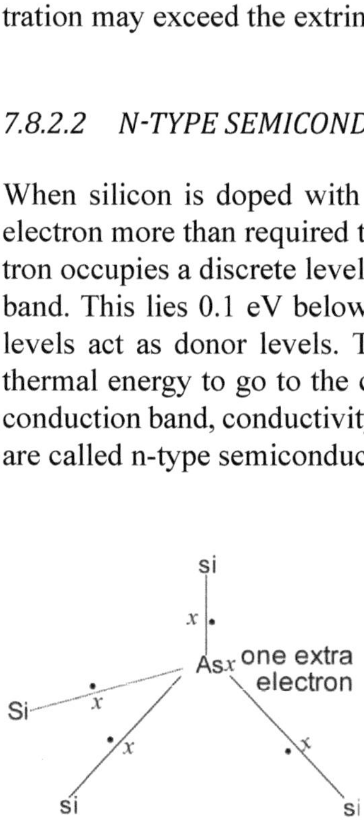

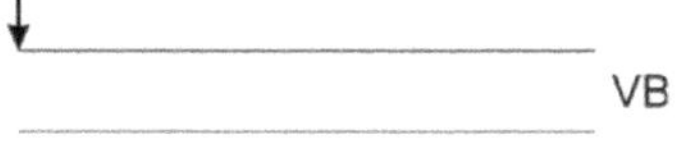

7.8.2.2.1 Temperature Dependence

The temperature dependence of carrier concentration and conductivity for a semiconducting material is shown in the figure given below.

At low temperatures, A, the carrier concentration is temperature dependent because of small gap of 0.1 eV between acceptor/donor states and the valence/conduction band is large that only very few electrons can make the transition when the temperature is increased, exhaustion occurs at B.

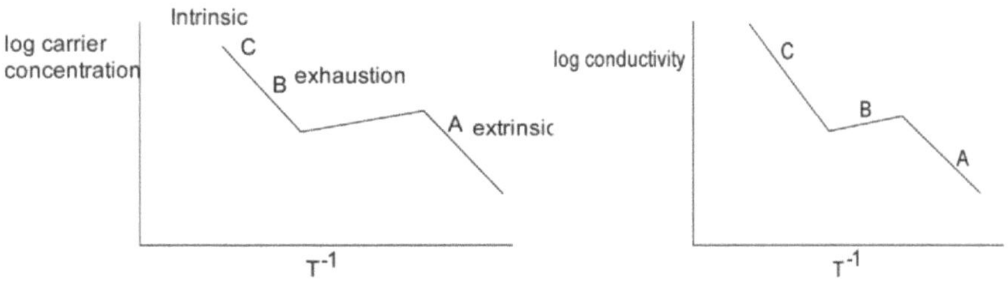

Extrinsic, exhaustion and intrinsic regions in a semiconductor.

When the concentration of extrinsic carriers attains its maximum value, the carrier concentration does not depend on temperature. At this stage, the conductivity shows a gradual decrease with rise in temperature.

KEYWORDS

- **extrinsic semiconductor**
- **Brillonin zone**
- **band theory**
- **electrical resistivity**
- **conduction band**
- **valence band**

CHAPTER 8

Electrical Properties

ABSTRACT

Thermoelectric effects arise at a junction of two different metals. When two different metals are in contact, electrons cross the junction from one metal to other giving rise to electric field. This effect is known as Peltier effect. The Hall effect and dielectric materials have been described in this chapter. Piezoelectric, ferroelectric and antiferroelectric crystals have also been discussed. A crystal may behave as piezoelectric without being ferroelectric. Optical processes such as photoluminescence, fluorescence and phosphorescence have been explained in detail with the help of energy diagram. When alkali halides are doped with activators such as thallous ion (Tl^+), luminescence is exhibited. Allowed transitions have also been explained. Dielectric materials are insulating materials used in capacitors and electrical insulators. Polarizability of dielectric materials has been incorporated in this chapter.

8.1 PELTIER EFFECT (METAL-SEMICONDUCTOR JUNCTIONS)

When two different metals are brought into contact, some electrons cross the junction from one metal to other and so electric field is established. Thus metal junction becomes a source of e.m.f. known as pelier e.m.f. of the order of a few millivolts.

Consider the junction between a metal and an *n*-type semiconductor. At this junction, the Fermi levels are at the same height on either side if the equilibrium is achieved. The regions near the interface on the semiconductor will be depleted of electrons and so assumes a positive charge and

Solid-State Chemistry: A Modern Approach. Ashok Kumar Jha, PhD (Auth.)

metal side becomes negatively charged. The positively charged region in the semiconductor side serves as a barrier against flow of electron. This region is known as barrier layer. This gives a potential difference V at the interface such that $eV = \phi_m - \phi_s$

Where ϕ_m and ϕ_s are the work functions of the metal and semiconductor, respectively. Work function of the metal becomes $\phi_m - \phi_e$, where ϕ_e is the energy difference between the Fermi level bottom of the conduction band.

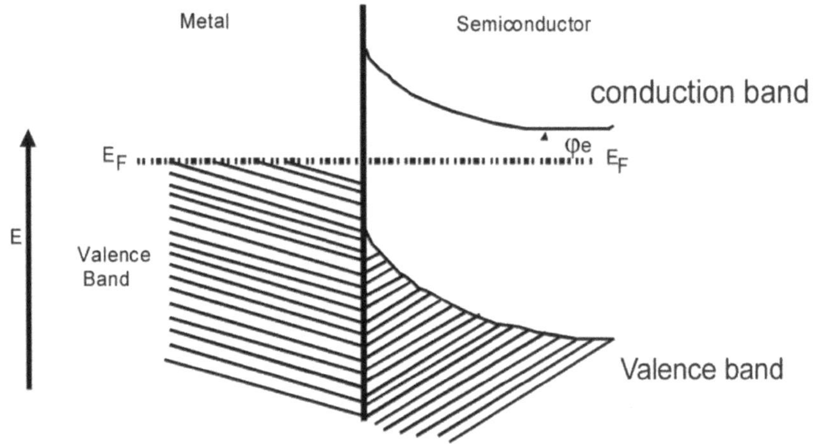

Band structure at metal-semiconductor junction

The Fermi level is below the bottom of a conduction band in a *n*-type semiconductor. So an energy $\phi_e + \frac{3}{2}\ K_BT$ is required for flowing electron from metal to conduction band of the semiconductor. $\frac{3}{2}\ KT$ is the extra amount of energy to have the kinetic energy of free electrons.

If an external voltage V is applied to make the semiconductor side positive with respect to metal, height of the barrier is increased to $\phi_m - \phi_s + ev$, and thus electron flow from semiconductor to the metal. If the external voltage is applied in the reverse direction, semiconductor side becomes negative with respect to metal, height of the barrier is reduced to $\phi_m - \phi_s - ev$. There results a net flow of current into the metal, which is the case of forward bias.

The net current during the forward bias may be expressed as

$$I = Io\left(\exp\frac{eV}{K_BT} - 1\right)$$

Where *Io* is the saturation current and

$$Io = \text{ABoT}^2 \exp - (\phi_m - \phi_s) / KT$$

A is the contact area, and *V* is the potential.

During the forward bias, the current increases exponentially.

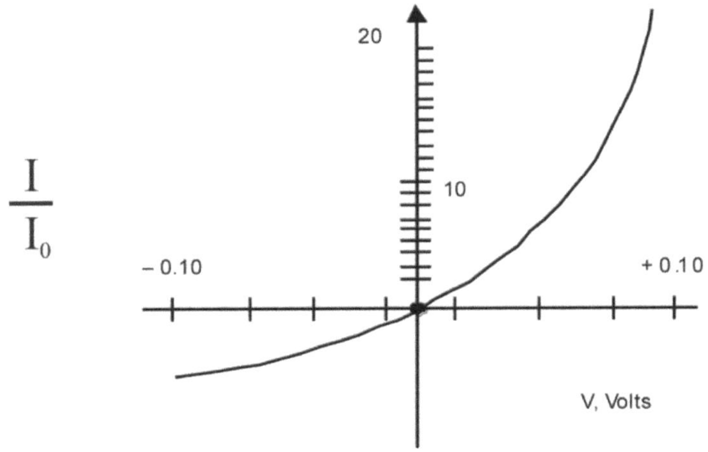

8.2 METAL–METAL JUNCTIONS

If two different metals are brought into contact, electrons flows from one to other due to the equilibrium tendency of Fermi levels. Metal 1 gets negatively charged and metal 2 becomes positively charged, which give rise to electrical double layer. A reverse electron current is generated from metal 1 to metal 2 at equilibrium and a contact potential develops between the surfaces of different metals. The electrical double layer is limited to the contact surface between the two metals.

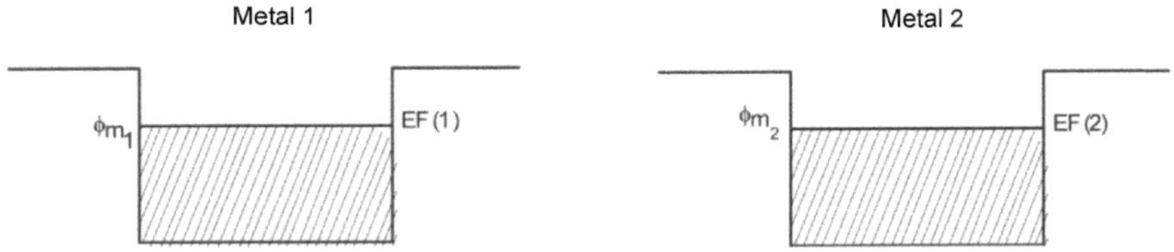

i) Development of contact potential between two different metals before contact

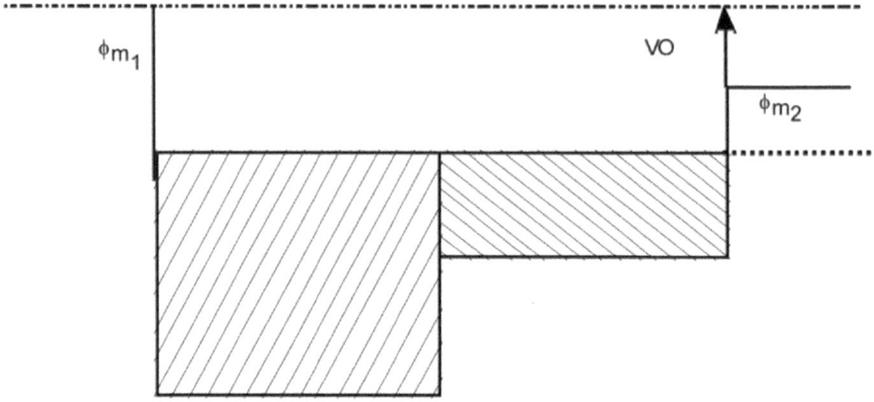

ii) Development of contact potential between two different metals at equilibrium after contact.

Let V_o be the contact potential

$$V_o.e = \phi_{m1} - \phi_{m2}$$

$$\text{or, } V_o = \frac{\phi_{m1} - \phi_{m2}}{e}$$

Value of V_o is of the order of a few tenths of a volt.

If two junctions between different metals are maintained at temperatures T_1 and T_2, a voltage V_{12} develops. This phenomenon is called Seebeck effect. V_{12} is known as Seebeck voltage, which can be expressed as

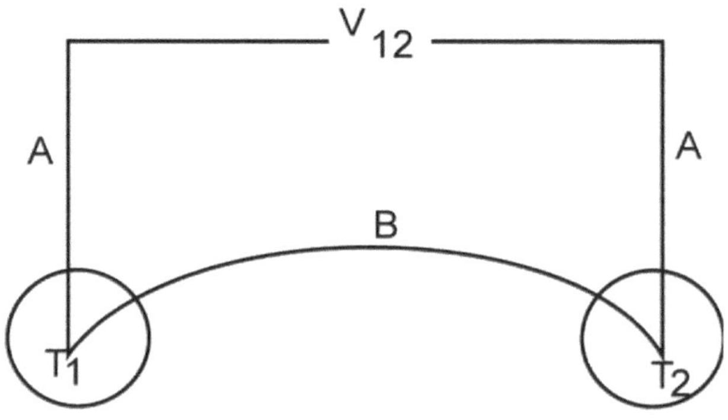

$\frac{dv12}{dT} = \alpha_{12}$ where α_{12} is Seebeck coefficient.

8.3 THE HALL EFFECT

If a current I (amperes) is allowed to flow through a solid in one direction and a magnetic field H is applied perpendicular to the direction of current flow, a potential difference develops across the sample perpendicular to H and I. Electrons follow curved path due to magnetic field H and so one end of the conductor becomes short of electrons than the other end. Thus unequal distribution of electrons takes place and a potential gradient is produced called Hall effect. When I and H are unity, the potential gradient for a material is the Hall coefficient. Let σ be the current density flowing through the solid, let it be the applied field and F_1 be the deflecting force F_1 due to it

$$F_1 = \text{He}\,\mu \qquad (8.1)$$

Where μ is the mobility of electrons and e their charge.

Let E be the Hall potential. Hall potential is an equal and opposing force F_2 to the deflecting force F_1

$$F_2 = \text{Ee} \quad = H\sigma\,\text{Re}$$
$$= \text{H ne}\,\mu\,\text{Re} \qquad (8.2)$$

Where $\sigma = \text{ne}\,\mu$

N—no of charge carriers

μ—mobility

From (8.1) and (8.2) σ-conductivity

$$\text{He}\,\mu = \text{Hne}\,\mu\,\text{R.e}$$
$$1 = \text{neR}$$
$$\text{Or, } R = \frac{1}{ne}$$

Hall coefficient has many practical applications. Most of metals have negative Hall coefficient, which show that electrons are current carriers.

In an intrinsic semiconductor, number of free electrons equals the number of holes. If the mobility is same, Hall voltage will be zero. But electrons are more mobile than positive holes and so a net Hall effect is produced. R also depends on temperature but is different for n- and p-type semiconductors.

In case of ionic conducting material, n and μ are low as a result of which Hall voltages cannot be detected. For a solid electrolyte $RbAgI_5$, Hall effect is considerable due to high ion conductivity of Ag^+ ion.

8.3.1 P-n JUNCTION

A junction between an *n*- and *p*-type silicon can be understood more easily. But *n*-type germanium and *p*-type gallium arsenide is a type of heterojunctions. Here behavior of the unction under external electric field has been studied. Electron flow from *n* side to *p* side when an *n*-type semiconductor comes in to contact with a *p*-type semiconductor. *p* side becomes more negatively charged and *n* side becomes positively charged. An electrical double layer is developed on junction at equilibrium.

On application of external voltage *p* side becomes positively charged which respect to *n* side. New holes are created in the *p* side, which move toward the junction. New electron created in the *n*-type semiconductor also move toward the junction. When these holes and electrons recombine at the junction, a large current flows called forward bias. If *p* side is negative with respect to *n* side, holes and electrons move in opposite direction and finally disappear.

i)

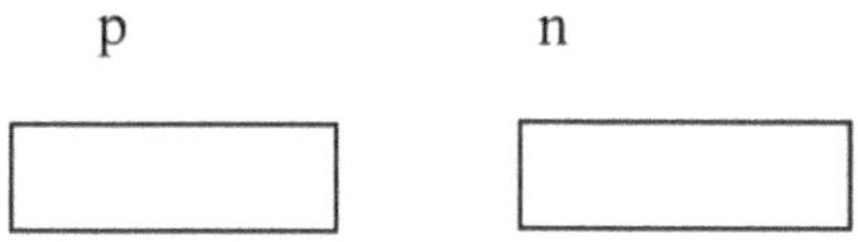

Semiconductor before contact

ii)

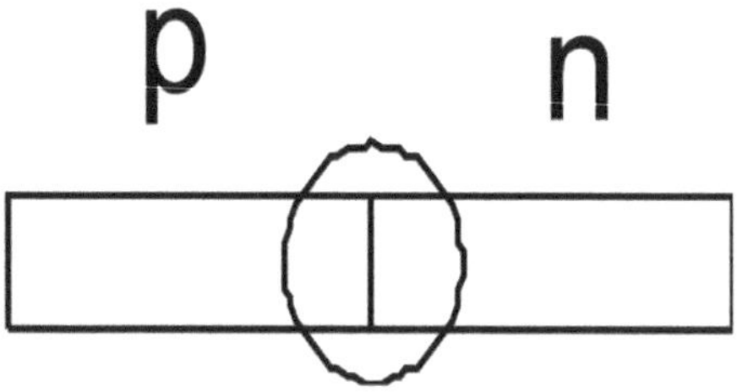

p-n Junction

iii)

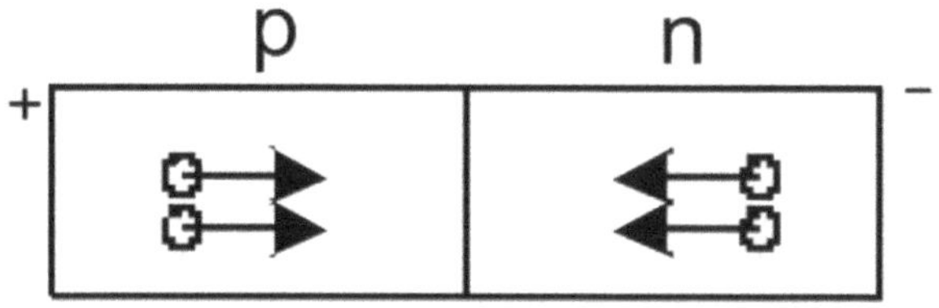

Forward bias

iv)

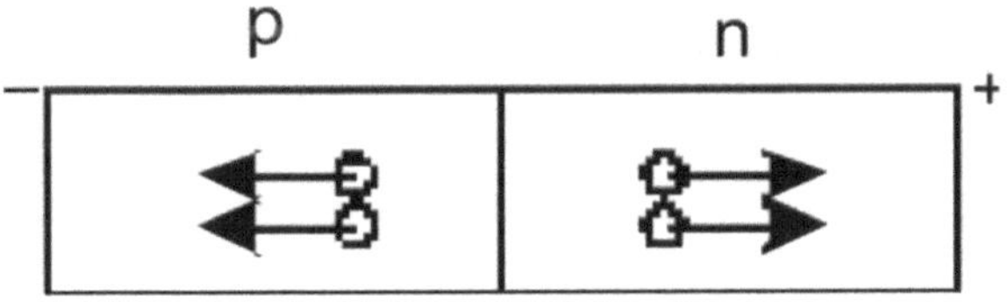

Reverse bias

The Fermi energy in the *n* semiconductor lies at a higher level, a flow of electrons takes place from *n* semiconductor to *p* semiconductor after contact of the semiconductors. At equilibrium, band bending occurs at the junction which opposes the flow of electrons from *n* side. The energy barrier is determined by the difference between the Fermi energies of *n* and *p* semiconductors.

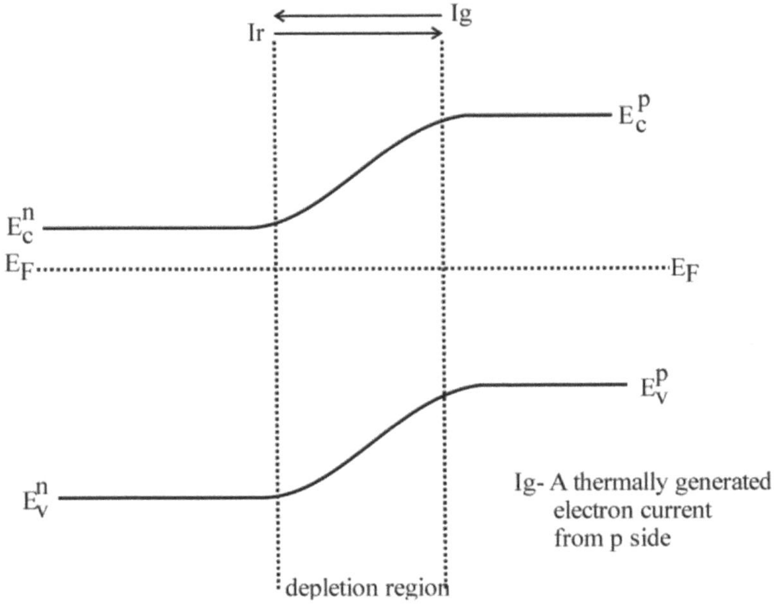

8.4 PHOTOVOLTAIC EFFECT

When photons are absorbed by *p–n* junction, electrons, and holes are generated. When these carriers diffuse to the junction, the built in electric field of the junction separates them at the energy barrier and as a result forward voltage across the barrier is produced.

Thus the appearance of a forward voltage across the junction is called the photovoltaic effect, which can deliver power to an external circuit. Application of this effect is that large area *p–n* junctions of *Si* are used to change solar photons to electrical energy.

8.5 DIELECTRIC MATERIALS

Dielectric materials are insulating material used in capacitors and electrical insulators. They should possess high dielectric strength without any degradation. The positive charges within the dielectric are displaced minutely in the direction of the electric field.

Thus when potential difference is applied, polarization of charge within the material occurs. When the voltage is removed, polarization disappears. Ferroelectric materials are those which retain residual polarization of charge even after removed of electric field.

By a pair of conducting plates, parallel to one another with vacuum between plates, dielectric properties may be explained. If the conducting plates are separated by a distance *d* and *A* is the area of the plates, ε_o is permittivity the capacitance co is defined as

$$C_o = \frac{\varepsilon_o . A}{d}$$

If a potential difference *V* is applied between the plates, a charge Q_o is stored on them

$$Q_o = C_o V$$

Here in the above figure dielectric substance is placed between the plates and same potential difference is applied. The amount of charge increases to Q_1 and s_0 capacitance becomes equal to C_1, The relative permittivity of the dielectric is given by

$$E' = \frac{C_i}{C_o}$$

Where E' depends on the degree of polarization. Relative permittivity of a ferroelectric material such as BaT_iO_3 is 10^3–10^4. Polarizability $\propto$ of the dielectric is defined as

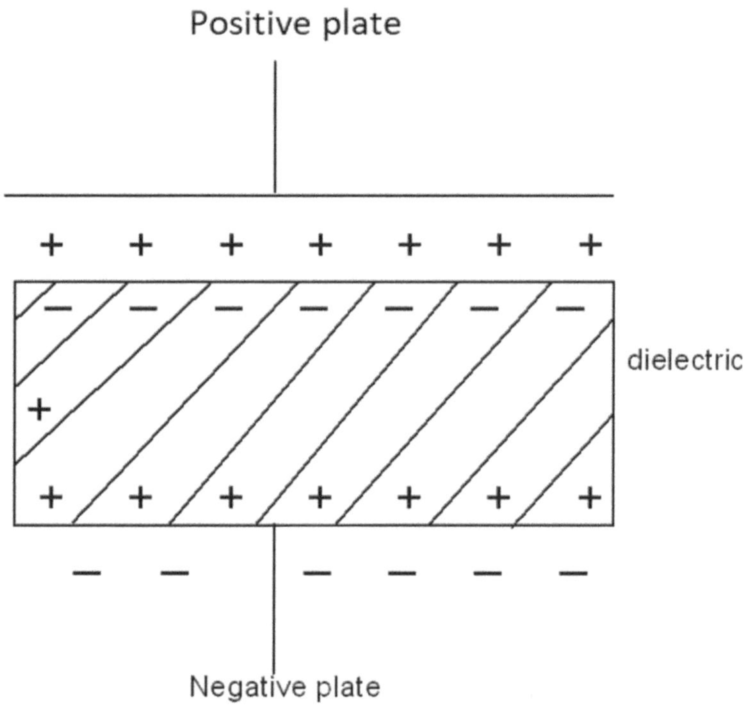

FIGURE 8.1 Dielectric material between the plates of a parallel plate capacitor.

$$P = \propto E$$

Where p is dipole moment included by electric field E

The polarizability can be considered of having four components

i) Electronic polarizability ($\propto_e$)
ii) Ionic Polarizability ($\propto_i$)
iii) Dipolar Polarizability ($\propto_d$)
iv) Space charge polarizability ($\propto_s$)

Polarizability $\propto$ can be given as

$$\propto = \propto_e + \propto_i + \propto_d + \propto_s$$

i) Electronic polarizability $\propto_e$ arises due to slight displacement of negatively charged electron cloud in an atom which takes place in all solids.

ii) Ionic polarizability is caused by separation of cations and anions in a solid. In ionic crystals, this is the main source of polarization.
iii) Dipolar polarizability arises in the substance, which has permanent electric dispoles that change their orientation with applied electric field and align accordingly.
iv) Space charge polarizability occurs in materials in which some long-range charge migration may occur. For example, cations move toward the negative electrode preferentially in NaCl having crystal defect. Obviously an electrical double layer is generated at the electrode-NaCl interface. The material starts behaving as a conductor then as a dielectric if such polarizability becomes considerable. It is quite clear that all materials do not show all types of polarization.

Polarization effects in a dielectric can be seen from the graph given below.

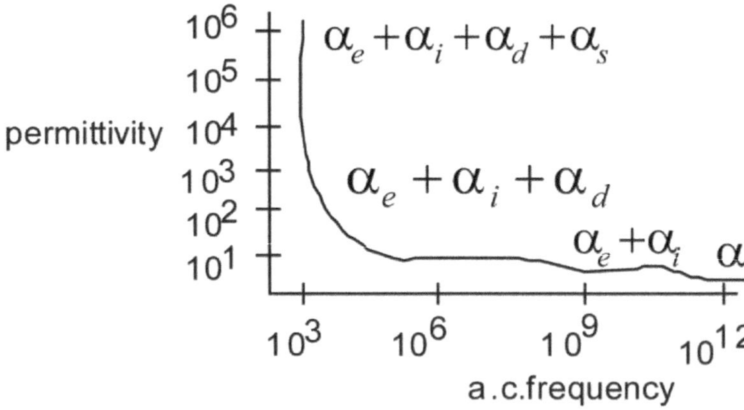

This graph has been obtained by making measurements over a wide range of a.c frequencies. At low frequency (10^3 Hz) total polarizability is equal to ($\alpha_e + \alpha_i + \alpha_d + \alpha_s$). At 10^9 Hz, polarizability is equal to ($\alpha_e + \alpha_i$) In good dielectric materials α_s and α_d contributions are absent, but only α_e and α_i contributions prevail. Permittivity $\in_\infty$ is composed of α_i and α_e only. Permittivity value $\in_\propto^1$ contain α_e contributions only.

Values of $\in_o^1$ and $\in_\propto^1$ for NaCl are 5.62 and 2.32 respectively.

Measurements are made over a range of frequencies, that is, audio frequency, radiofrequency, and microwave regions.

If a low-frequency sinusoidal e.m.f. is applied across a dielectric, all polarizations can occur before the field reverses. Such polarizations transmit energy through the dielectric. If the energy losses due to the joule heating are zero, energy is transmitted through the sample without dielectric losses.

8.6 PIEZOELECTRICITY

When a stress *Z* is applied to the crystal, electric polarization is changed. Thus it can be said that piezoelectricity is a phenomenon associated with induction of polarization under mechanical stress. Ferroelectricity is due to the spontaneous polarization even in the absence of external electric field. So spontaneous polarization can be induced by varying the temperature. If the polarization is reversed by the application of external electric field, the phenomenon is called ferroelectricity. In one dimensional notation the photoelectric equations are

$$P = Z_d + E_x \tag{8.3}$$

$$E = Z_S + E_d \tag{8.4}$$

Where P is the polarization, Z is stress, d is the piezoelectric strain constant, E is the electric field, X is the dielectric susceptibility, e is the elastic strain, and s is the elastic compliance constant. Eqs (8.3) and (8.4) exhibit the development of polarization by an applied stress and elastic strain by an applied electric field A crystal may behave as piezoelectric without being ferroelectric. For example, quartz is piezoelectric but not ferroelectric

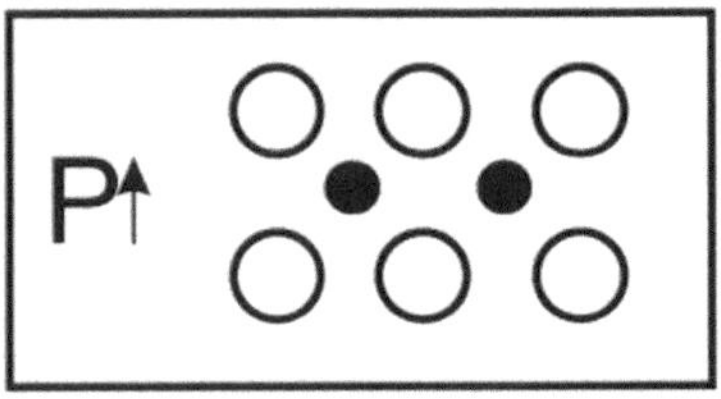

(a) Ferroelectric crystal without stress

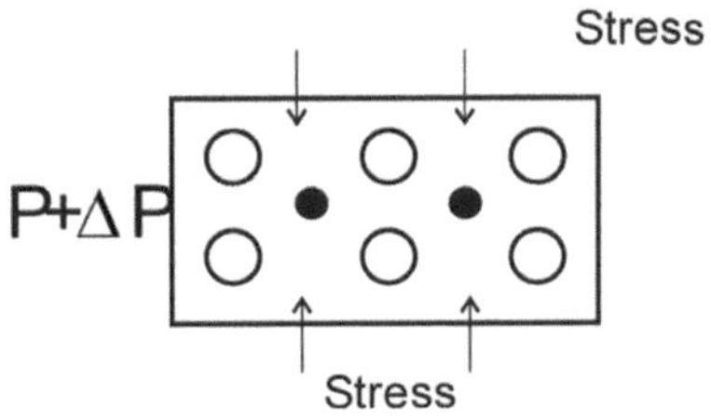

(b) Stressed ferroelectric crystal where polarization is changed by Δ P.

The lead zirconate lead titanate system is used in polycrystalline form with high piezoelectric coupling. The synthetic polymer polyvinylidenfluoride (PVF_2) is five times more piezoelectric than crystalline quartz. Stretched films of PVF_2 is five times more piezoelectric than crystalline quartz. Stretched films of PVF_2 are flexible and so used in medical applications.

8.7 FERROELECTRICITY

A ferroelectric crystal exhibits an electric dipole moment even without external field. In some crystals, the ferroelectric dipole moment does not change even by an electric field of maximum intensity. But on change of temperature, a change in the spontaneous moment is observed. Such crystals are known as pyroelectric. Lithium niobate $LiNbO_3$ is pyroelectric at room temperature and has a transition temperature of 1480 K.

Ferroelectricity disappears after a certain temperature called transition temperature. The crystal is said to be in a paraelectric state above the transition temperature.

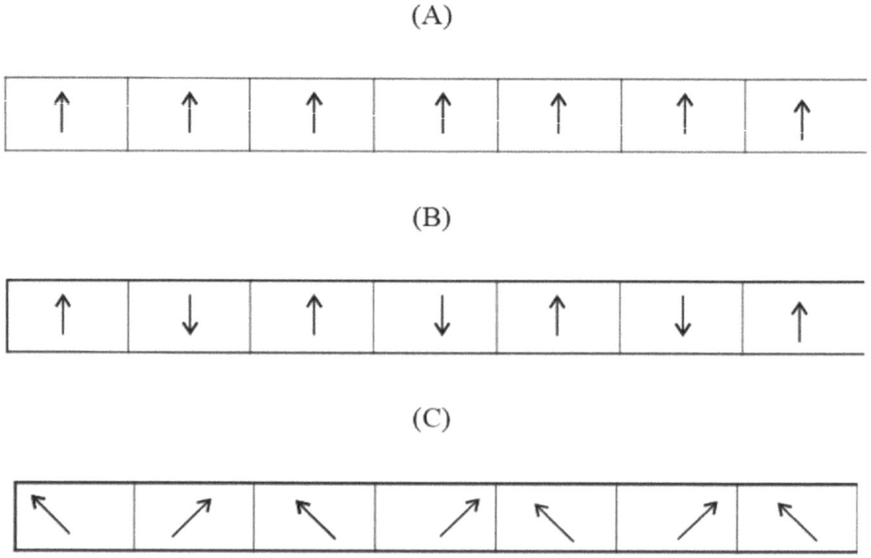

FIGURE 8.2 Dipole orientation is (a) ferroelectric (b) an antiferroelectric (c) a ferroelectric.

$BaTiO_3$ is an example of ferroelectrics. In this domain, structures form due to alignment of TiO_6 dipoles parallel to each other. The net polarization of a ferroelectric material is the vector resultant of the polarizations of the individual domains.

A necessary condition for ferroecletricity is that it should be at low temperature. If the temperature is raised, domain structure is destroyed. A large number of oxides having distorted perovskite structure are ferroelectric materials. $BaTiO_3$ and $PbTiO_3$ are ferroelectric whereas $CaTiO_3$ is not. Ferroelectric materials are used in capacitors due to high dielectric constants.

8.8 ANTIFERROELECTRIC MATERIAL

Spontaneous polarization also occurs in antiferroelectric materials. In antiferroelectric materials, individual dipoles align antiparallel to adjacent dipoles and net spontaneous polarization is zero. Above the antiferroelectric Curie temperature, the materials become paraelectric, for example, lead Zirconate ($PbZrO_3$), Sodium niobate ($NaNbO_3$), and ammonium dihydrogen phosphate ($NH_4H_2PO_4$).

Curie temperatures Oxides	Curie temperature
$PbZrO_3$	233°C
$NaNbO_3$	638°C
$NH_4H_2PO_4$	(−125°C)

The antiferroelectric state is a nonpolar one and no hysteresis loop occurs.

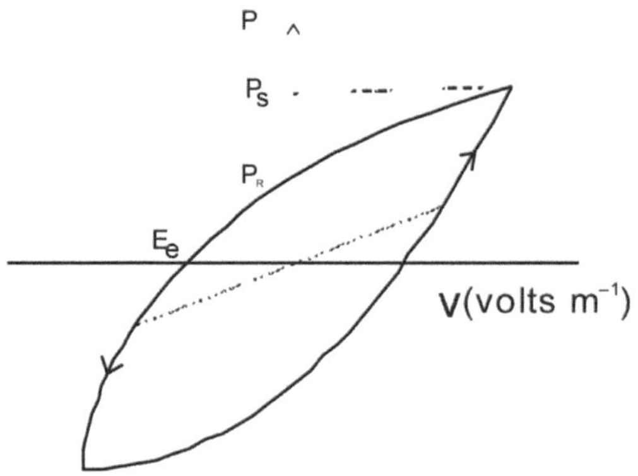

Hysteresis loop of a ferroelectric. Dashed line passing through the origin represents normal dielectric materials. Hysteresis loop can be understood as increase in field strength domains grow at the expense of others and at high field strength, polarization reaches a maximum. When the field is decreased, polarization adopts another path and polarization is retained even at zero field. This is called remanent polarization. Remanent polarization becomes zero when the field is applied in the opposite direction. The field requires to reduce the polarization to zero is called the coercive force. Further increase in field results in saturation at the polarization in the reverse direction and thus hysteresis curve is obtained.

Ferroelectric Curie temperature can be described as

$$\in_r = \frac{C}{T - T_c} + \in_{r_o} \text{ (Curie-Weiss law)}$$

Where T_c is Curie point and $\in_{r_o}$ is a constant having a value of the order of 1% or less then $\in_r$ which is due to the electronic polarizability only

$$\text{Thus } \in_r = \frac{C}{T - T_c}$$

8.9 PYROELECTRICITY

Pyroelectric crystals show spontaneous polarization, *Ps* that cannot be reversed by an applied electric field. *Ps* can be expressed as

$$\Delta Ps = \pi . \Delta T \text{ where } \pi \text{ is pyroelectric coefficient.}$$

On heating thermal expansion changes the sizes of the dipoles, for example, ZnO. Structure of ZnO contains a hexagonal close-packed array of O^{2-} ions with Zn^{2+} ions located in one set of tetrahedral sites. The ZnO_4 tetrahedral possesses a dipole moment and so the crystal has a net polarization. Polar impurity molecules are absorbed onto the crystal. One of important conditions for a crystal to be pyroelectric is noncentrosymmetric. The absence of a center of symmetry is noncentrosymmetric. Of the 32 point groups, optical activity is confined to 20 noncentrosymmetric point groups. Thus a piezoelectric crystal must be one of the point groups among the 20 active groups. So it is better to know space group before going into the investigations of pyroelectricity. Thus the absence of a center of symmetry is an essential requirement for the presence of pyroelectricity and piezoelectricity in a crystal.

Tow tetrahedra related by a center of symmetry can be shown as given below

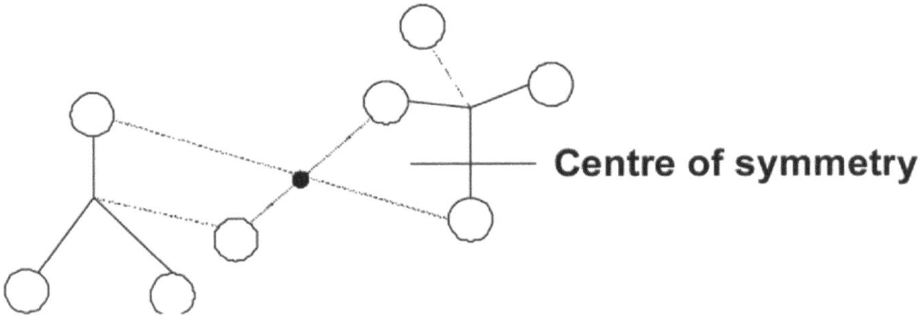

8.10 CENTROSYMMETRIC

The effect of any reflection or inversion operation is to convert a left-handed object into a right-handed one. Though individual tetrahedral do not possess a center of symmetry, groupings of tetrahedral possess a center of symmetry.

8.11 OPTICAL PROCESSES

Luminescence may be defined as the emission of light by a material after absorbing energy. A number of excitation sources are available. Photo-luminescence uses photons or light, electroluminescence uses electrical energy input. Fluorescence and phosphorescence are also the outcome of absorption of light. The incident radiation causes an increase in the electrical conductivity due to excitation of electrons from valence to conduction band. Electron thus excited into the conduction band returns to the valence band. If a crystal absorbs energy and emits it back in the visible range, the phenomenon is called luminescence. If the re-emission is instantaneous, the phenomenon is called fluorescence. If it takes a longer time more than 10^{-7} s, it is called phosphorescence. Luminescence is influenced by the presence of imperfections and additional energy states are introduced by them within the band gap of a semiconductor. Such states are called recombination centers and traps. A recombination center is a discrete energy state near the center of the band gap. Other localized levels are called traps, which are capable of capturing either an electron or hole.

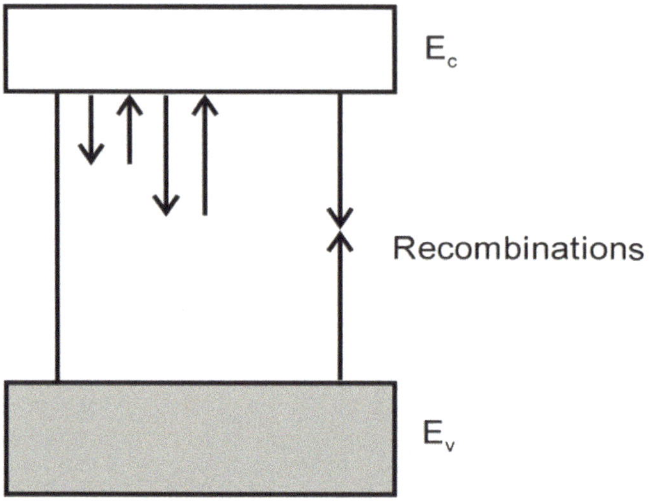

Transitions in an impure semiconductor

The presence of recombination centers or traps gives rise to several kind so possible transitions.

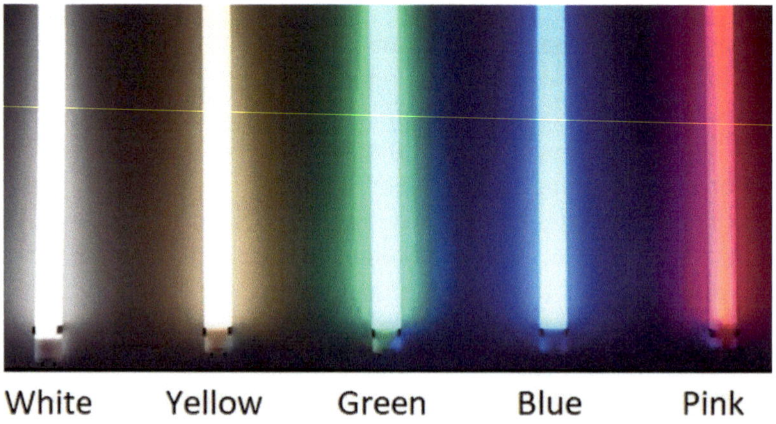

8.12 CO-ORDINATE MODEL

Total energies of the ground and the excited electronic states of the center as a function of configuration co-ordinate may be seen below

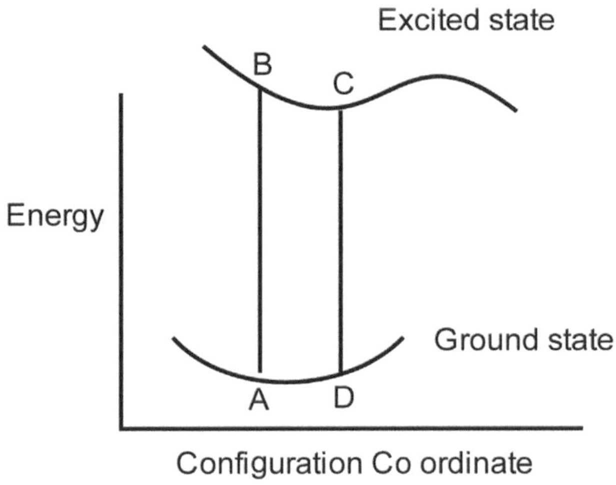

FIGURE 8.3 Luminescence.

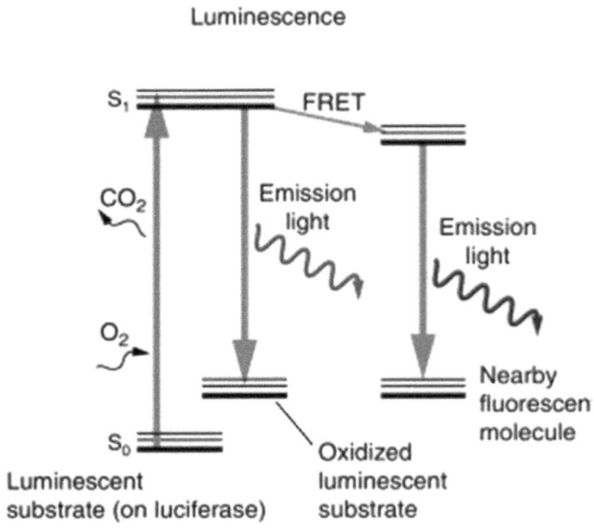

FIGURE 8.4 Diagram of luminescence.

During absorption, an electronic transition takes place from A to B that is vertical. It takes place so fast that the momentum and the position of the ion are unchanged. In the excited state, the system relaxes from B to C. The energy differences at B and C is released as phonons, that is, lattice vibrations. Emission of photon then takes place from C to D which is luminescent emission. The system may again relax from D to A and so the frequency of luminescent photon is less than that of the incident photon.

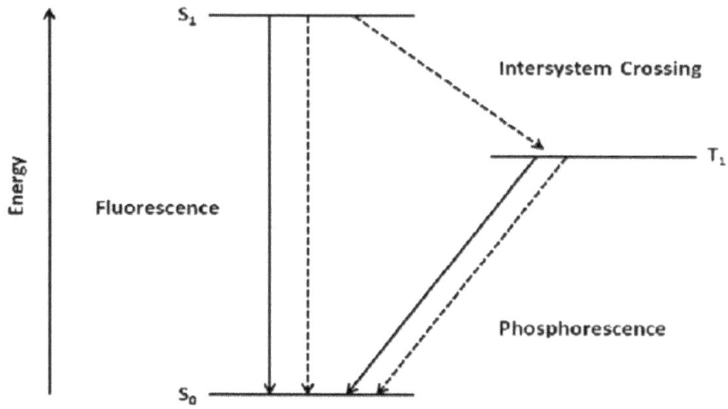

FIGURE 8.5 Energy diagram of phosphorescence.

One possibility is that movement of ions in the excited state of the center may result in a disposition of the ground and excited state. Equilibrium position C in excited state may correspond to higher vibrational level in the ground state (point D). In such cases, a radiation less transitions occur. So it has become clear that absorption of a radiation may not always emit radiation.

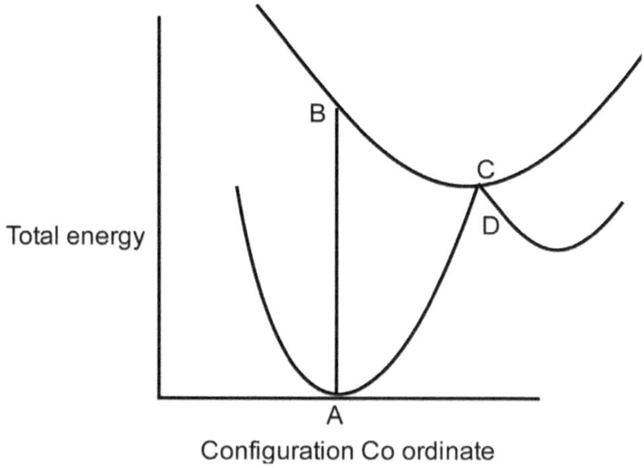

FIGURE 8.6 Radiation less transition.

8.13 LUMINESCENCE IN ALKALI HALIDES

Alkali halides are colorless and absorb in the U.V. region. When alkali halides are doped with activators such as thallous ion (Tl^+) ion, luminescence is exhibited, for example, doped KCl. Thallous ions replace potassium K+ ions. There is a six-field co-ordination of Cl^- ions for each Tl^+ ion. In cubic symmetry, the allowed transitions are $1_{A1} \rightarrow 1_{T1}$ and $1_{A1} \rightarrow 3_{T1}$ having bands at 233 and 282 nm, respectively. Additional bands at 250, 252, 290, and 293 nm are observed for higher dopant concentrations of Tl^+ ions.

8.13.1 LUMINESCENCE IN ZNS TYPE PHOSPHOROUS

ZnS and CdS are the materials in which the activator ions are (Li^+, Na^+, Cu^+, Ag^+, Au^+). These ions have a closed outermost shell electronic configuration. If Ag^+ ions enter the ZnS lattice, The Ag^+ ions are monovalent and Zn^{2+} ions are divalent. So one sulfide ion (S^{2-}) vacancy for every two silver ions is introduced. If a Cl^- ion substitutes for a S^{2-} ion, cation vacancies are generated. Al^{3+} and Ga^{3+} ions are also used as coactivators. It is worth mentioning that phosphors, for example, ZnO, ZnS, CdS, and ZnS are all direct gap semiconductors. If diamagnetic or paramagnetic impurities are introduced, additional energy states develop within the band gap. These energy states are responsible for luminescence.

Various color ranges can be obtained when concentration of Cu, Ag, Au and Al, Ga, In in ZnS are varied. Important luminescences obtained are red copper, blue copper, green copper, and blue self activated copper. Thus zinc sulfide in presence of activators and co-activators can exhibit different kinds of luminescence. The origin of luminescence in red Cu and blue Cu is due to possible electronic transitions from the conduction band to a localized state above the valence band. The transition may also take place from the valence band to a localized state below the conduction band.

In the transition between localized states in the band gap does not involve either valence band or conduction band.

When Mn^{2+} ions enter the ZnS structure, a bright orange luminescence results due to d–d transitions. d–d transition is not allowed in the free ion state but-becomes prominent when Mn^{2+} ions find position in the lattice of ZnS. If Fe^{2+}, Co^{2+}, or Ni^{2+} are introduced to ZnS lattice, green luminescence is destroyed. This is due to the presence of re-combination centers in the band gap and so such ions capable of destroying luminescence are known as killer ions.

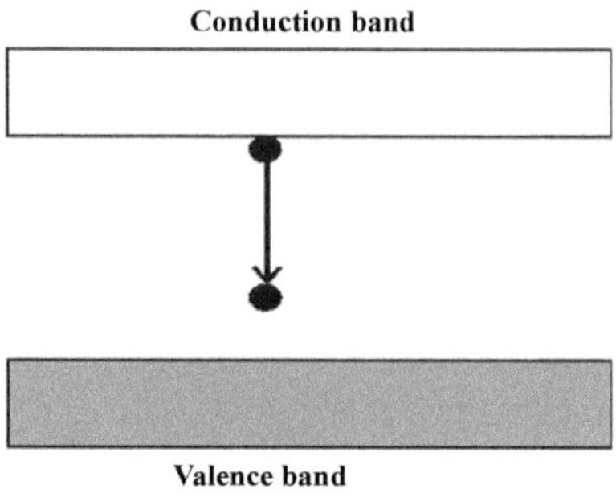

(a) Electron transition

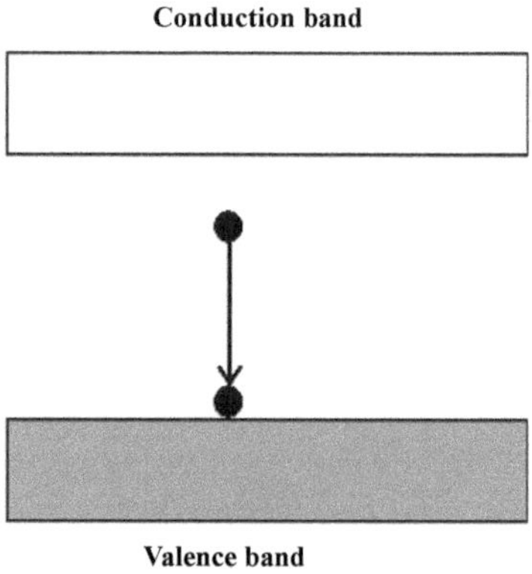

(b) Hole transition

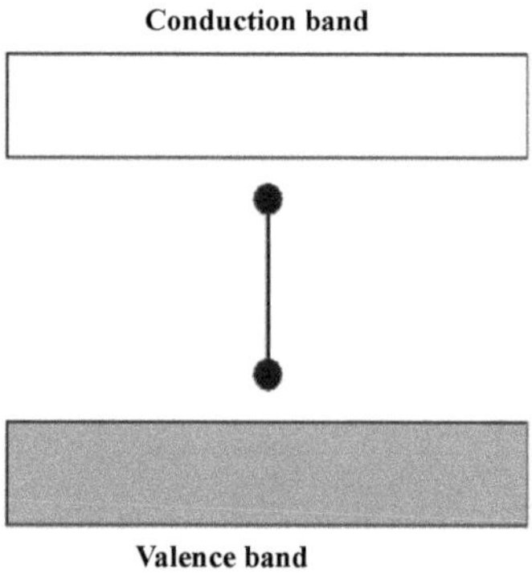

(c) Transition between localized states within the band gap

When fluorapatite is doped with Sb^{3+} ion, it fluoresces blue and when doped with Mn^{2+} it gives orange-yellow fluorescence. Trivalent europium (Eu^{3+}) is an activator ion used in red phosphors for color television screen. In YVO_4: Eu^{3+}, the vandate group absorbs energy in the cathode ray tube but the emitter is Eu^{3+}. Color television screens require three primary cathodoluminescent colors, for example.

YVO_4: Eu^{3+} ....red
ZnS: Ag^+ ....blue
ZnS: Cu^+ ....green

8.14 ANTISTOKES PHOSPHORS

Antistokes phosphors are those in which photons of higher energy are emitted than that of the incident exciting radiation. Thus it is possible to convert infrared radiation to visible light of higher energy. In normal luminescence phenomenon process of excitation takes place in two stages as shown below (Figure)

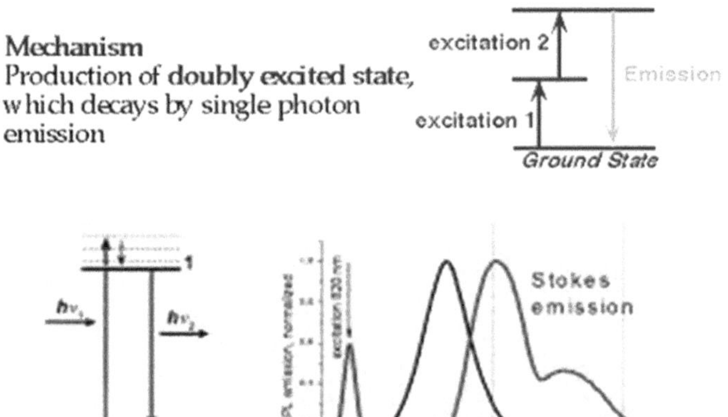

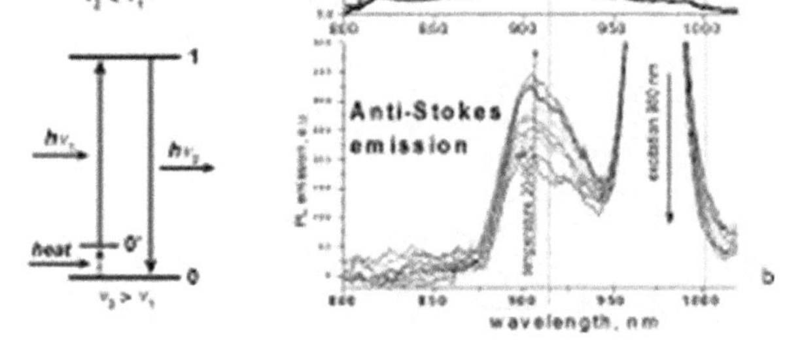

The best studied antistokes phosphors are the host structure YF_3, La $(WO_4)_2$, which have been doped with Yb^{3+} as a sensitizer and Er^{3+} as an activators. These materials are capable of converting infrared radiation into green luminescence.

8.15 LASER AND MASER ACTIONS

Laser stands for light amplification by stimulated emission of radiation and Maser stands for microwave amplification by stimulated emission of radiation. Lasers are used as oscillators and amplifiers used in microsurgery and masers are used as low noise amplifiers in microwave communication systems. Population of states of atoms, ions, or molecules over the allowed energy levels can be expressed by Maxwell–Boltzmann distribution law as

$$n_i = n_o \exp\frac{\{-(E_i - E_o)\}}{K_B T}$$

where n_i—number of atoms, ions or molecules in excited energy level, n_o—number of atoms, ions or molecules in the ground state, K_B = Boltzmann constant, and T—absolute temperature.

The excited states are thin populated in comparison to the ground state. When an electromagnetic radiation falls on ions, ions are excited due to adsorption of photon. Now the excited ion decays to the ground state and photon is emitted which is called spontaneous emission. But the excited ions may decay to an intermediate level known as met stable state having larger lifetime.

KEYWORDS

- **Peltier effect**
- **piezoelectric**
- **ferroelectric**
- **fluorescence**
- **phosphorescence**
- **luminescence**

CHAPTER 9

Magnetic Properties

ABSTRACT

Magnetic behaviour is shown by transition metals and inner transition metals. Effect of temperature on magnetism has been studied. Paramagnetic substances obey the curie law at high temperature. Curie point and neel point have been explained with the help of diagram. Paramagnetic, ferromagnetic and antiferromagnetic behaviour have also been shown in the diagram. Calculations of magnetic moments and experimental determination of magnetic susceptibility have been discussed in this chapter. Proper explanation of quenching of orbital contribution towards magnetic moment has been given.

9.1 INTRODUCTION

Inorganic solids exhibit magnetic effects. Magnetic behavior is exhibited by transition metals, lanthanides as well as actinides because the presence of unpaired d and f electrons. The unpaired electrons may be oriented on atoms and paramagnetism develops. If domains are arranged to be parallel, material possesses an overall magnetic moment to give rise to ferromagnetism. If they are aligned in antiparallel, overall magnetic moment is zero and hence antiferromagnetic. If the alignment of domain is antiparallel and unequal, the behavior is ferrimagnetic.

9.2 MAGNETIZATION VECTOR (I)

Magnetic poles are present as pairs of opposite poles, which are different from electric charge. Such a pair of opposite poles is called magnetic

Solid-State Chemistry: A Modern Approach. Ashok Kumar Jha, PhD (Auth.)

dipole. If a current i is passed through a loop of area A, magnetic dipole is generated in the direction perpendicular to the plane of the current flow.

Magnitude of the dipole

$d = i.\,A$

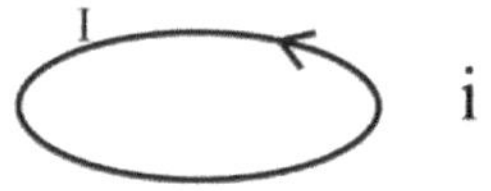

A is area of loop
d is a vector directed from the negative to positive pole
Magnetic moment μ is defined as
$\mu = md$ where m is a unit dipole.

Atoms have electrons that move in close orbits giving rise to the magnetic moment.

When a substance is put in a magnetic field H, density of lines of force in the sample is known as magnetic induction B.

$$B = H + 4\pi \frac{I}{H}$$

Where I represents the magnetic moment of the sample per unit volume P is permeability and K is susceptibility.

$$\frac{B}{H} = 1 + 4\pi \frac{I}{H}$$

$$P = 1 + 4\pi k$$

Where $K = \frac{I}{H}$

The molar susceptibility x is given by

$$x = \frac{KF}{d}$$

F represents formula weight and d density.

If $P < 1$ and K, x are small and slightly negative, substance is diamagnetic. If $P > 1$ and k, x are positive, substance is paramagnetic. As a result, paramagnetic substance is attracted by a magnetic field whereas diamagnetic substances experience repulsion.

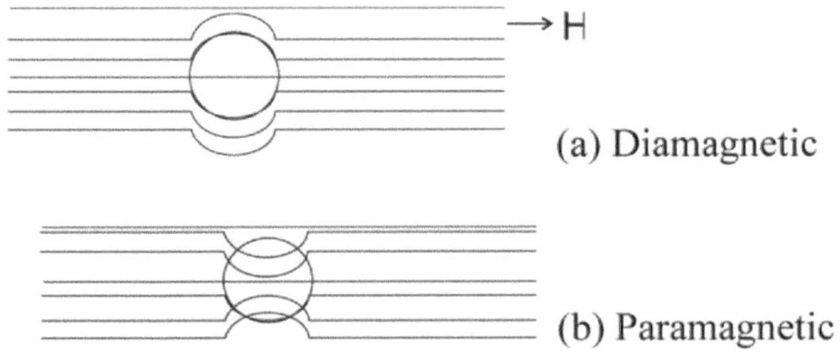

9.2.1 EFFECT OF TEMPERATURE

Many paramagnetic substances obey the curie law at high temperature. According to this law, magnetic susceptibility is proportional to temperature

$$x = \frac{C}{T} \tag{9.1}$$

Where C represents the curie constant. Eq (9.1) can be written as

$$x = \frac{C}{T + \theta} \tag{9.2}$$

Where θ is the Weiss constant. Eq (9.2) is known as Curie–Weiss law. When x^{-1} is plotted against T; the following graph is obtained.

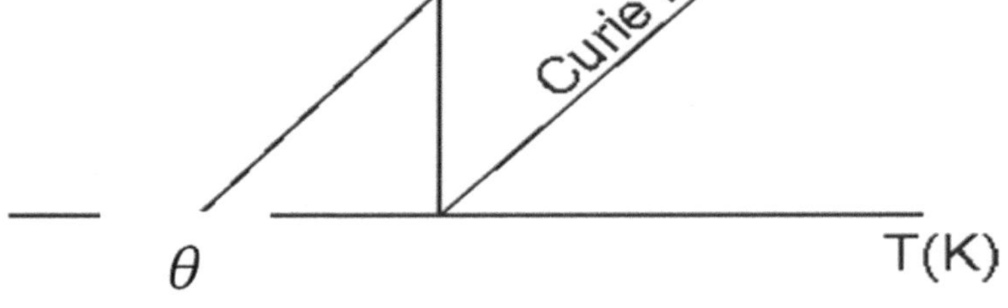

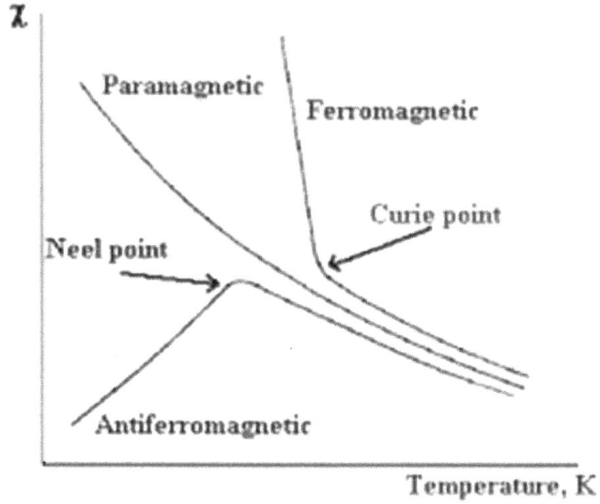

Curie and Curie–Weiss law is not a fit for ferro and antiferromagnetic substance.

Greater susceptibility is shown by ferromagnetic material at lower temperature. Susceptibility decreases with rising temperature. Above a certain temperature that is Curie temperature, material does not behave as ferromagnetic but behaves as paramagnetic. In case for ant ferromagnetic substance, the value of x increases with rising temperature up to a critical temperature is known as Neel point T_N. Above this point, paramagnetic behavior is shown by the material.

When unpaired electrons are present in the material and align in a magnetic field, paramagnetism is developed. In ferromagnetic materials, the electron spins are aligned parallel due to the interaction of spins in the crystal structure. Large values of x show parallel alignment of spins. In antiferromagnetic materials, the electron spins are aligned antiparallel to have small x values. In case of paramagnetic substance, the thermal energy of ions and electrons acts opposite to applied magnetic field. Thermal energy and magnetic field partially neutralize the effect of each other. So in case of paramagnetic substance x decreases with increasing temperature.

9.3 LANGEVIN EQUATION

When an alteration in the flux through electrical circuit takes place, an induced current is set up to oppose the flux change. As long as field remains present, the induced current remains in an electron orbit within an atom. The magnetic field resulting from induced current acts opposite to the applied field.

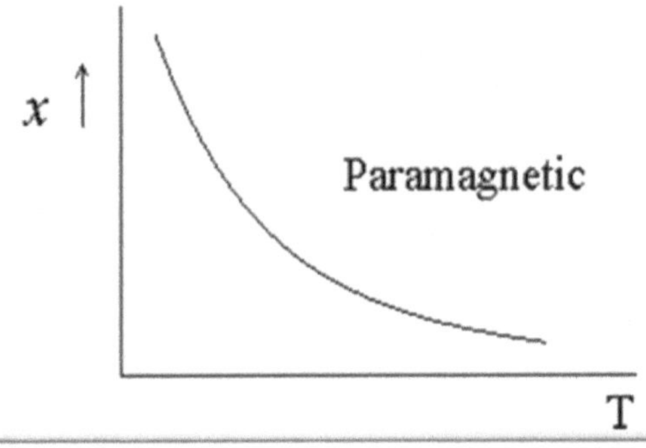

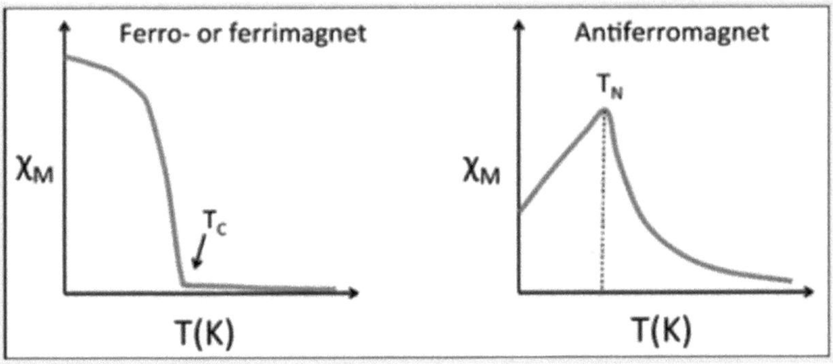

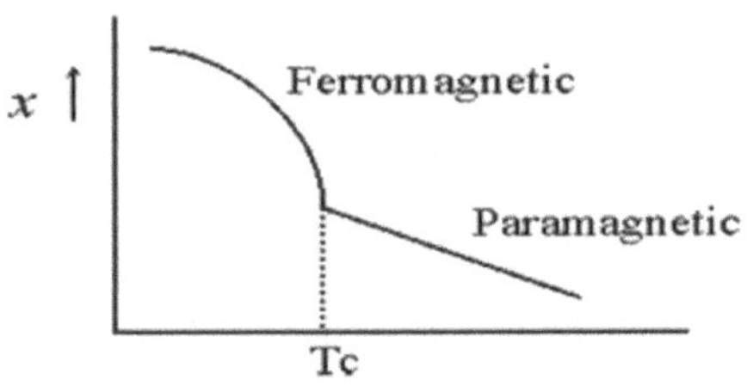

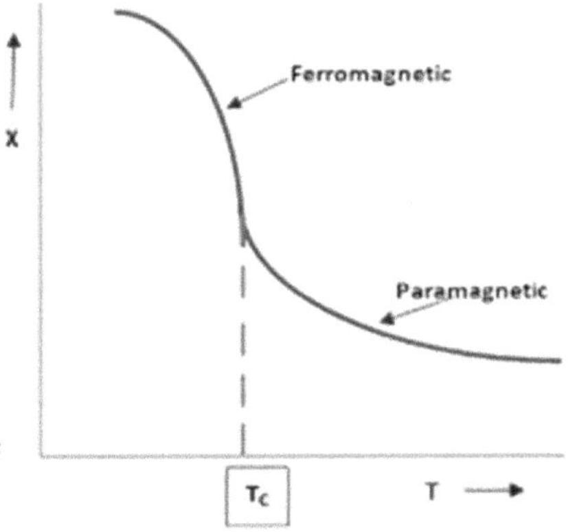

Figure 1

T_N – Neel point

The applied field makes the magnetic dipole arising of the motion of the electron in the orbit to precess in a course reverse to the course of electron motion. So let us consider that the electron orbit is perpendicular to the field direction

$$w = \frac{eB}{2m} \ (S.I) \tag{9.3}$$

Where w represents angular velocity and B represents magnetic field intensity.

When the average electron current around the nucleus is zero, a finite current around the nucleus will be generated on the application of magnetic field.

The Larmer precession of Z electrons is equivalent to an electric current

$$I = \text{charge} \times \text{revolutions per unit time}$$

$$= -\text{Ze}\left(\frac{1}{2\pi} \ \frac{eB}{2m}\right) \tag{9.4}$$

Where revolutions per unit time $= \dfrac{w}{2\pi}$

The product of current and area of the loop provides the magnetic moment μ of a current loop

$$\mu = I \times \pi\rho_2$$

$$\mu = \frac{Ze^2 B}{4m} \rho^2 \tag{9.5}$$

Here $\rho^2 = x^2 + y^2$, which is the mean square of the perpendicular distance of the electron from the field axis through the nucleus.

The mean square distance of the electrons from the nucleus is

$$r^2 = x^2 + y^2 + z^2 \tag{9.6}$$

If the charge distribution is spherical and symmetrical

$$x = y^2 + z^2$$

From (9.6) $r^2 = 3x^2$ and $p^2 = 2x^2$ or $x^2 = \dfrac{P^2}{2}$

Let x be the diamagnetic susceptibility per unit volume and N be the number of atoms per unit volume

$$x = \frac{N\mu.}{B}\mu o$$

$$x = \frac{NZe^2.B}{4m.B}.\rho^2.\mu o$$

$$x = \frac{-NZe^2}{4m}.\frac{2}{3}r^2.\mu o$$

$$x = \frac{-NZe^2}{6m}.r^2.\mu o$$

$$x = \frac{-\mu oNZe^2}{6m}.r^2 \quad (9.7)$$

This is the classical Langevin result. It has been concluded that diamagnetic susceptibility is a negative quantity show by substances. Its value is independent of temperature and field. Therefore the diamagnetic contribution of the ion cores for dielectric solids is described by the Langevin diamagnetism equation.

9.3.1 CALCULATION OF MAGNETIC MOMENTS

The electron may be regarded as a small sphere negative charge spinning on its axis. Thus an electron traveling in a close path around a nucleus produces magnetic moment. It can be said that magnetic properties of an atom or ion results from a combination of spin moment of electron and orbital moment. The unit of magnetic moments of atoms ions and molecules is Bohr magneton. Bohr magneton is defined as

$$1\ \text{BM} = \frac{eh}{4\pi mc}$$

Where e = electron charge

h = Planck's constant
m = electron mass
c = velocity of light

The spin-only formula for an electron is

$$\mu_{S.O} = g\sqrt{s\ (s+1)}$$

Where s represents the spin quantum number $\frac{1}{2}$ and g is the gyrometric ratio

If the atoms or ions that contain more than one unpaired electron, spin-only formula is provided by

$$\mu_S = g\sqrt{s\,(s+1)}$$

Where S is the sum of the spin quantum number of unpaired electrons

In addition to this, motion of an electron around the nucleus may give rise to orbital moment and so orbital moment also contributes to the total magnetic moment expressed as

$$\mu_{S+L} = \sqrt{4s(s+1)+L(L+1)}$$

If the orbital contribution is quenched

$$\mu_{S.O} = \sqrt{4s(s+1)}$$

$$= \sqrt{4.\frac{n}{2}\left(\frac{n}{2}\right)\left(\frac{n}{2}+1\right)}$$

$$= \sqrt{4n(n+2)}\text{B.M}$$

Where n is no. of unpaired electrons.

9.3.2 QUENCHING OF ORBITAL CONTRIBUTION

According to quantum theory, angular momentum component L_z and square of the total orbital angular momentum L^2 remains fixed in a central field. L_z will no longer be fixed but L^2 may be a constant. If L_z averages to zero, orbital angular momentum is quenched. When the spin–orbit interaction takes place, the spin may pull some orbital moment along with it. When orientation of the spin and orbit magnetic moments are in parallel direction, the value of g will be greater than 2. The experimental results agree with theoretical variation of sign of the spin–orbit interaction.

When the 3d subshell is more than half filled $g > 2$ when the subshell is half filled, $g = 2$ and when the subshell is less than half filled $g < 2$.

Crystal field has an important role in the quenching process. Crystal field splits the degenerate levels into nonmagnetic levels. Energy separation in this is greater than $\mu.H$. Magnetic field has a small perturbation than crystal field. Suppose the ground state of an ion having one p electron. It is triply degenerate $m_1 = \pm 1, 0$ in the free atom. But the energy of the ion is lowered if the free atom comes in contact with surrounding, that is, lattice. As a result of this noncubic potential develops in place of cubic potential. This effect is seen in Mn^{3+} and Cu^{2+} ions, alkali and silver halides.

9.4 EXPERIMENTAL DETERMINATION OF MAGNETIC SUSCEPTIBILITY BY GOUY'S METHOD

In this method, a thin glass tube filled with the sample has been vertically suspended from the beam of a balance. The lower end to the tube lies between the poles of a electromagnet. The weight of the sample is known with and without field.

Let dF be the small force, x—volume susceptibility, dv—small volume, H—magnetic field

$$dF = H.x.dv.\frac{dH}{dx} \text{ Where } \frac{dH}{dx} \text{ is the field gradient.}$$

If A is the cross sectional area and dx-small height of the sample

$$A.dx = dv$$

$$dF = H.x.A.dx.\frac{dH}{dx}$$

$$= H.x.A.\ dx$$

On integration

$$\int_{HO}^{H} dF = \int_{HO}^{H} H.x.A.dH$$

or, $F = A.x.\frac{1}{2}(H^2 - HO^2)$

Where H and HO are the lower and upper limits of magnetic fields.

$$x_g = \frac{x}{d}$$

$$F = A.x_g.\text{d}.\frac{1}{2}(H^2 - H_O{}^2)$$

$$\text{Density(d)} = \frac{\text{mass}}{\text{volume}} = \frac{m}{Axl}$$

or, $d = \frac{m}{Al}$

$$A.d = \frac{m}{l}$$

$$F = xg.\frac{m}{l}.\frac{1}{2}(H^2 - H_O{}^2)$$

If H_O is neglected

$$x_g = \frac{2l \times F}{H^2 \times m}$$

$$x_g = \frac{2l \times \text{change in weight of the sample}}{H^2 \times \text{weight of the sample}}$$

or, $x_m = \frac{2l \times \text{change in weight of the sample}}{\text{weight of the sample}} \times \text{Molecular weight}$

The value of x_m is corrected for the diamagnetism. Corrected value of x_m is used to calculate the effective magnetic moment with cur equation.

$$\mu\text{eff} = {}^{2.83}\sqrt{x_m^{\text{corr}}.(T - \theta)}\text{BM}$$

The wave functions of the three *p* orbital are given as *Zf*(*r*), *xf*(*r*), and *yf*(*r*) corresponding to P_z, P_x, and P_y orbitals. In the crystal system, electron cloud is supposed to be close to positive ions. When the electric field has no axial symmetry, all the three wave functions give different energies

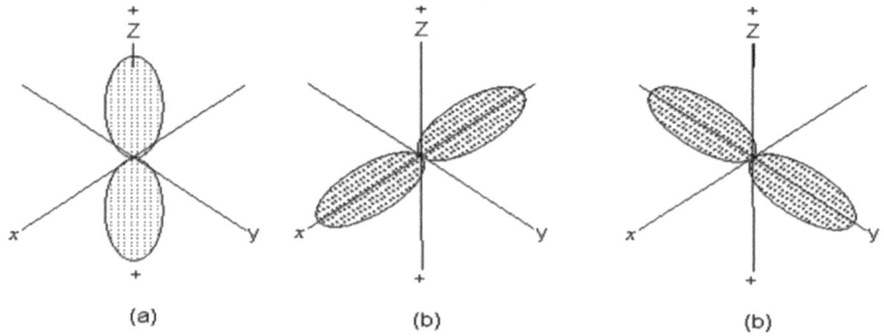

All the three wavefunctions $zf(r)$, xf(r), and $yf(r)$ referring to P_z, P_x, and P_y, respectively, are degenerate ion free atom but the charges surrounding them disturbs the degeneracy in the crystal p_x and p_y becomes degenerate and P_z becomes lower in energy.

-------------- ---------------------p_x, p_y

---------------------p_z

(c) Free atom (d) on application of electric field having no axial symmetry

Thus the three eigenstates in the crystal field are p functions in which angular lobes are directed along x, y, z axes. When orbital moment of the each of the states becomes zero, the effect is known as quenching. In this way quenching may be explained in terms of quantum theory.

9.4.1 MAGNETIC PROPERTIES OF TRANSITION METAL OXIDES

The transition metal oxides, for example, MnO, FeO, and NiO are paramagnetic at high temperatures and at ordering in magnetic structures takes place at low temperature. In such type of oxides, *d* electrons are localized on metal ion, which give rise to magnetic properties.

Many important magnetic oxides have spinel structure. The spinels may be shown as

$$\text{Normal } [A]^{\text{Tet}}\ [B_2]^{\text{Oct}}\ O_4$$

$$\text{Inverse } [\text{B}]^{\text{tet}}\ [\text{A}, B]^{\text{Oct}}\ O_4$$

Tet. stands for tetrahedral void and oct. stand for octahedral void.

The parent spinel is $MgAl_2O_4$ having cubic lose packed structure in which Mg^{2+} occupies tetrahedral interstices and Al^{3+} occupies octahedral one. Several spinel structures have been reported till now. Sulfides, selenides, and tellurides are also spinels. Different cations may be introduced into spinels as given below.

Mg_2TiO_4

$LiAlTiO_4$

$Li_{0.5}Al_{2.5}O_4$
$LiNiVO_4$
Na_2WO_4

In inverse spinels, half of the B ions occupy tetrahedral and half of B ions and the A ions occupy octahedral sites. Some of inverse spinels have been given below

$MgFe_2O_4$ $MgIn_2S_4$
$NiFe_2O_4$ Mg_2TiO_4
$MgIn_2O_4$ Zn_2TiO_4

Ferrites of the type MFe_2O_4 are magnetic spinels. Here M may be Fe^{2+}, Ni^{2+}, Cu^{2+}, and Mg^{2+} Fe^{3+} ions having d^5 configuration have no crystal field stabilization energy in an octahedral site.

$$d5 - t_{2g}{}^3 e_g{}^2,\ \text{C.F.S.E} = -3 \times \frac{2}{5}\Delta_O + 2 \times \frac{2}{5}\Delta_O = 0$$

Larger divalent ions go preferentially to octahedral sites and Fe^{3+} is distributed over tetrahedral and octahedral sites both. These ferrites are all either antiferromagnetic or ferromagnetic. Ions on the tetrahedral sites have magnetic spins that are antiparallel to the ions on the octahedral sites. $ZnFe_2O_4$ is a normal spinel that is antiferromagnetic at low temperature because the Fe^{3+}ions couple antiparallel. Coupling between tetrahedral and octahedral sites is very weak and hence T_N is low for $MgFe_2O_4$ a residual magnetic moment develops and so it is ferromagnetic. Spinel may be inverse because Fe^{3+} ions are both on tetrahedral and octahedral sites. Experimental values indicate that $MgFe_2O_4$ transforms to the normal spinel structure at high temperature.

Manganese ferrite $MnFe_2O_4$ is a 80% normal and 20% inverse spinel. This has been found ferromagnetic having a magnetic moment of ~5 B.M. Mixed ferrites can be where M may be Mg, Ni, Co, Fe, Mn, etc. If $x = 0$, these ferrites are largely inverse

$$\left[Fe^{3+}\right]^{tet}\left[M^{2+}, Fe^{3+}\right]^{oct} O_4$$

Zinc ferrite $ZnFe_2O_4$ is a normal spinel at room temperature. Alignment of spins of Fe^{3+} ions occupying octahedral sites of $ZnFe_2O_4$ does not take place but are random. So $ZnFe_2O_4$ is paramagnetic.

KEYWORDS

- **paramagnetic**
- **ferromagnetic**
- **Curie point**
- **Neel point**
- **quenching**

CHAPTER 10

Phase Transition, Phase Rule, and Diagrams

ABSTRACT

Phase transition is one of the important aspects of solid-state chemistry. Thermodynamic criteria of phase change have been discussed with reference to Gibbs free energy, Helmholtz free energy, internal energy and entropy with the help of the diagrams. Structural changes accompany phase transition. Phase rule, component and degree of freedom have been discussed with reference to one component and two component systems. Simple eutectic system such as lead–silver system has been explained with the help of phase diagram. Phase diagram of binary systems with solid solution has also been dealt with. Peritectic system of two solid solution has been shown by the phase diagram. Phase transition may be attributed to reconstructive and displacive transitions. Critical size of nuclei in nucleation and growth transition along with Avrami plot have been discussed.

10.1 INTRODUCTION

Phase equilibrium and phase diagrams explain the transformation from one crystal structure to another. Phase diagrams are actually the graphical forms for a certain range of temperature and pressure under thermodynamic equilibrium. Temperature, pressure, and concentration are the factors that determine the phase. Phase transitions are shown by a variety of systems in technological regions. Phase diagrams are

Solid-State Chemistry: A Modern Approach. Ashok Kumar Jha, PhD (Auth.)

explained by the phase rule derived by Gibbs thermodynamic quantities like entropy, free energy, and heat capacity have also to be discussed during the phase rule.

10.2 THERMODYNAMIC CRITERIA

A solid system exists as a homogeneous phase or a mixture of several phases. Different forms of solid are interconvertible subject to temperature and pressure change. On heating a homogeneous solid, it may change to another phase with unaltered composition. But the crystal structure changes. In some transformations, structure and composition both change.

Gibb's free energy is given as $G = E - TS + PV$

Where G is the Gibbs free energy, E, T, S, P, and V are the internal energy, temperature, entropy, pressure, and volume of the solid.

Let the change in Gibbs free energy for solid phases 1 and 2 be in equilibrium at constant temperature and pressure

$$G_1 - G_2 = [\,(E_1 - TS_1) + PV_1\,] - [\,(E_2 - TS_2) + PV_2\,] = 0$$

There is almost no change in volume for a solid. So $P\,(V_1 - V_2) = 0$

$$G_1 - G_2 = E_1 - E_2 - T\,(S_1 - S_2)$$

$$\Delta G = \Delta E - T\Delta S$$

$$\Delta G = \Delta A,$$

where $\Delta A = \Delta E - T\alpha S$

A is known as Helmholtz free energy. Thus, it is clear that the form of solid with lower free energy will be stable in phase transition at 0K or lower temperature, phase with lower internal energy is stable. At increased temperature, the structure with increased entropy is favored. Suppose a solid exists in two forms α and β at 0K, the α phase will be more stable if $E_A < E_B$ and $A_\alpha < A_\beta$.

If $S_B > S_\alpha$, S_{BT} will be increased faster than $S_\alpha T$. The two free energy curves will intersect at some temperature, at this temperature, the two phases α and β will coexist. Above this temperature, phase β becomes stable. This temperature is known as transition temperature, phase α transforms to phase β.

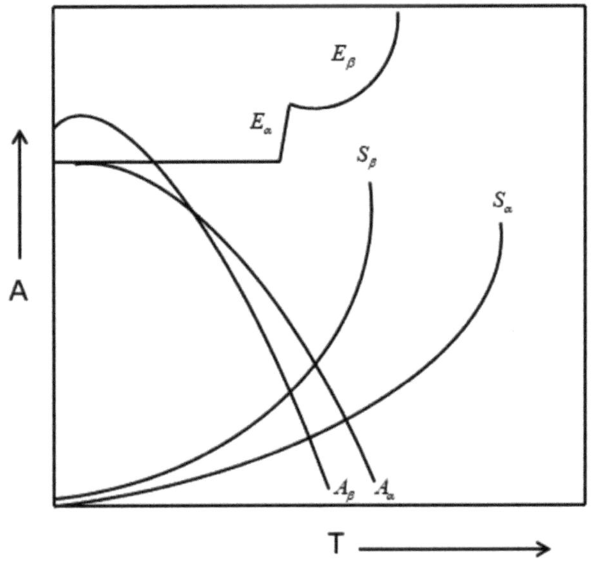

FIGURE 10.1 Variation of A, E, and S with temperature.

The temperature at which $A_\alpha = A_\beta$ is said transition temperature.

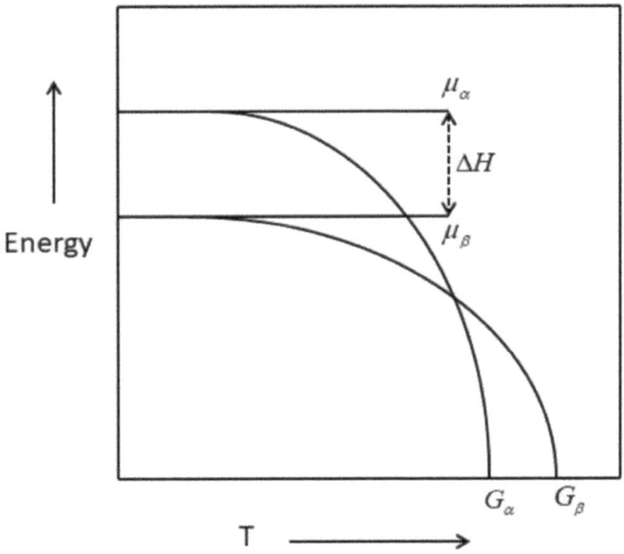

FIGURE 10.2 Changes in Gibbs free energy and enthalpy with temperature in a phase transition.

The relation between Gibbs free energy and enthalpy is given as follows:

$$G = H - TS$$

The change in enthalpy and Gibbs free energy in α and β phase transition is given as:

$$G_\alpha - G_\beta = H_\alpha - H_\beta - T(S_\alpha - S_\beta)$$

Let us consider a binary solid solution AB and such that $E_{AA} > E_{BB}$, where E_{AA} and E_{BB} are the interaction energies between A–A and B–B, respectively.

If a small amount of B is added to A, free energy is lowered. On continuous addition of the constituent B, a solid solution of A and B is formed with all possible compositions.

Thus, when B is added to A, free energy decreases and then increases. If A is added to B, free energy decreases and then increases. The minima occurs at mole fractions $(x_B)_1$ and $(x_B)_2$. At mole fraction $(x_B)_1$, α form is stable and at $(x_B)_2$, β form is stable. In the region between $(x_B)_1$ and $(x_B)_2$, the two phases α and β both exists.

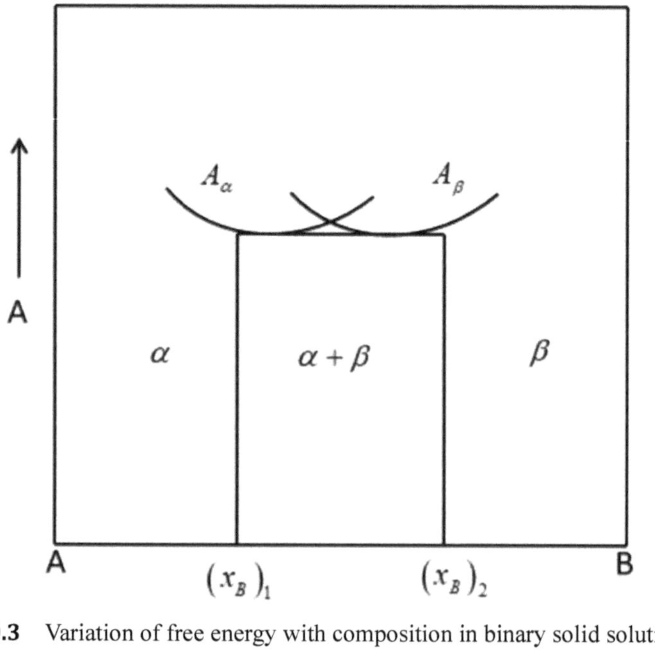

FIGURE 10.3 Variation of free energy with composition in binary solid solution.

The phase transition has also been described by the Clausius–Clapeyron equation.

The Gibbs free energy change with respect to temperature and pressure is given as follows:

$$\left(\frac{\delta G}{\delta T}\right)_P = -S, \quad \left(\frac{\delta G}{\delta P}\right)_T = V$$

$$\frac{dP}{dT} = \frac{\Delta S}{\Delta V} = \frac{\Delta H}{T\Delta V}$$

If α and β are the two phases of solid

$$\frac{dP}{dT} = \frac{\Delta H_t}{V_\beta - V_\alpha}$$

where T represents the transition temperature at which two forms α and β have molar volumes V_α and V_β. These molar volumes are in equilibrium. ΔH_t represents the molar heat of transition. This is the thermodynamic condition for equilibrium between two solid modifications.

The thermal transition temperature of AgI is 427 K at which hexagonal form changes to cubic. At 750 k CsCl (b.c.c. structure) changes to NaCl (f.c.c. structure). Pressure transitions also take place and thus structure gets changed. At 89 K.bar pressure structure of ZnO changes from wurtzite to NaCl.

10.3 SIO_2 SYSTEM

Silica is one of the most common oxide abundant in earth's crust. The polymorphism of silica on heating at atmospheric pressure can be shown as

$$\alpha-\text{quartz} \xrightarrow{573^\circ\text{C}} \beta-\text{quartz} \xrightarrow{870^\circ\text{C}} \beta-\text{tridymite} \xrightarrow{1470^\circ\text{C}}$$
$$\beta-\text{cristobalite} \xrightarrow{1700^\circ\text{C}} \text{liquid}$$

At a pressure of 900 atm tridymite form disappears, whereas cristobalite field disappears at 1600 atm. Above this pressure (1600 atm), the only stable polymorph exists. The increased pressure generally produces polymorphs having a higher density. It has been found that above 90,000–120,000 atm, one polymorph of stishovite is stable. In addition, many metastable polymorphs of SiO_2 also exist.

The densities of SiO_2 polymorphs are given below.

Polymorph	Density(gm per cc)
Tridymite	2.298
Cristobalite	2.334
Quartz	2.647
Stishovite	4.28

10.4 PHASE TRANSITION WITH STRUCTURAL CHANGES

From the data of density and crystal, it has become clear that changes in structure take place during phase transition. During thermal transition, high-temperature form has higher symmetry. But in the case of transition caused by pressure, high-pressure form is closely packed having a smaller volume. This can be understood very clearly by the example of NaCl-type and CsCl-type solid. When pressure is applied, NaCl-type solid changes to CsCl-type solids. In NaCl type structure, coordination number is 6, whereas in CsCl the coordination number is 8. Thus, on applying pressure coordination number changes from 6 to 8. When CsCl-type solid is heated, NaCl-type solid structure is obtained. The coordination number changes from 8 to 6.

In AB_2 type solid, for example, TiO_2, CaF_2, a distorted perovskite is transformed to the cubic structure at high temperature. Structural phase transition can also be noticed in mercuric iodide crystal. HgI_2 undergoes structural change when $\alpha - HgI_2$ transforms to $\beta - HgI_2$ at 127°C. Cuprous chloride (CuCl) also shows a similar phase transition.[1]

The lattice of α – HgI_2 becomes a good host for organic impurities as a result of the change in bonding and structure. The intercalation here takes place for impurities and lattice defects and physical properties thus changing. Such intercalation with organic compounds is also possible with clay or clay minerals. Some of them worth mentioning are smectite group of minerals, illite, and kaolinite.

$AgNbO_3$ and $AgTaO_3$ also undergo structural phase transition when they are cooled from the cubic perovskite phase.[2] At room temperature, weak ferroelectric behavior is shown by $AgNbO_3$.

AgI shows a first-order structural phase transition at 420 K from the hexagonal wurtzite structure of β phase.[3] There are three phases: α, β, and

γ. In the α – phase, I^- ions occupy a bcc lattice. β phase has hexagonal wurtzite structure, whereas γ phase has cubic sphalerite structure. β phase is metastable with respect to the γ phase. Thus, structural phase transitions are seen in silver iodide solid. During the study of compressed hydrogen, structural phase transition has been observed. In compressed hydrogen, metallization takes place as a result of band overlapping in molecular phases.[4] As the density increases, hydrogen undergoes structural phase transition as given below.

$$\text{HCP} \rightarrow \text{BCC}$$

$BaTiO_3$ and $SrTiO_3$ also show structural phase transition. At $T_c = 110$ K, $SrTiO_3$ exhibits antiferro transitions in which the number of formula units in the unit cell changes.[5] $BaTiO_3$ exhibits ferro transition in which a number of formula units in the unit cell do not change.

The temperature and pressure are the two variables that affect and cause phase transition. Two phases can coexist in a single system when the transition of a substance takes place from one phase to another.[6] Pressure and temperature both are used for the transformation of one phase to another. This phenomenon of phase change can be explained with the help of a phase rule and diagram.

10.5 PHASE RULE

The phase rule has been derived thermodynamically and given by the equation:

$$F = C - P + 2$$

where P is the number of phases present in equilibrium, C is the number of components, and F is the number of degrees of freedom. The degree of freedom means the number of independent variables, such as temperature, pressure, and concentration.

10.5.1 PHASE

The phase is defined as physically distinct and separable by a distinct boundary in a system. The simplest example in the case of water has been cited. It has three phases: solid, liquid, and gas. In the case of solids, the crystalline phases exist. The example of magnesium silicate minerals can be shown. Enstatite, $MgSiO_3$ and forsterite, Mg_2SiO_4 are the minerals of magnesium but both are in different phases.

Solid phases occur as solid solutions. A solid solution of $\alpha - Al_2O_3$ and Cr_2O_3 occurs but their phases are different. Both of them form solid solutions at high temperatures and they may be changed to one phase.

In the case of solids, different crystalline phases can be obtained with the same chemical composition known as polymorphism.

10.5.2 DEGREE OF FREEDOM

The degree of freedom of a system is the number of independent variable factors, such as temperature, pressure, and concentration to define the system completely.

In the case of equilibrium between water and vapor, the only variable is pressure. So the degree of freedom is 1. If a system is at equilibrium in three phases, such as ice, water, and water vapor, this can be realized at a certain temperature and pressure. If pressure or temperature varied, all liquid changes to solid. If the temperature is raised, solid phase, that is, ice disappears. Here there is no independent variable. Thus, at equilibrium between three phases of water, that is, ice, liquid, and vapor, the degree of freedom is zero.

$$F = C - P + 2, \quad F = 1 - 3 + 2 = 0$$

When a saturated solution of salt is in contact with the salt, either temperature or pressure is considered to define the system completely. In case, the system has only one degree of freedom.

If we take a solution of a salt, temperature or pressure and concentration of the salt are considered. Obviously, the two variable factors are here temperature and concentration or pressure and concentration. Thus, the degree of freedom is 2.

10.5.3 COMPONENT

The component may be defined as the minimum number of constituents by means of which the composition of each phase is represented.

In water system, water consists of three phases, that is, solid, liquid, and vapor. But all three phases have only one constituent that is water. All three phases can be represented by H_2O and thus water system is a one-component system.

In the case of dissociation of NH_4Cl, there are two phases: solid and gas. This will be a two-component system.

$$NH_4Cl \rightarrow NH_3(g) + HCl(g)$$

Decomposition of calcium carbonate may be shown as:

$$CaCO_3 \rightleftharpoons CaO(s) + CO_2(g)$$

Only two constituents are required to define the system and so decomposition of calcium carbonate is a two-component system.

10.6 PHASE EQUILIBRIUM

If a two-component system exists in two phases α and β in equilibrium and dn_1 moles of one component get transformed from phase α to β, the chemical potential of the given component remains the same in every phase.

Let μ_1 be the chemical potential:

$$\mu_1(\alpha) = \mu_1(\beta)$$

10.6.1 ONE-COMPONENT SYSTEM

In a one-component system, pressure and temperature are the variables.

From the phase rule ($F = C - P + 2$)

$F = 2$ if $C = 1$, $P = 1$. If two phases are present, $F = 1 - 2 + 2 = 1$

Thus for a one-component system, phases are two polymorphs, α and β, liquid and vapor. Each phase has a region separated from the other phase which can be seen in Figure 10.4.

BE is the transition curve for polymorph α and β. The transition temperature changes with pressure. FC is the change of melting point of polymorph β with pressure.

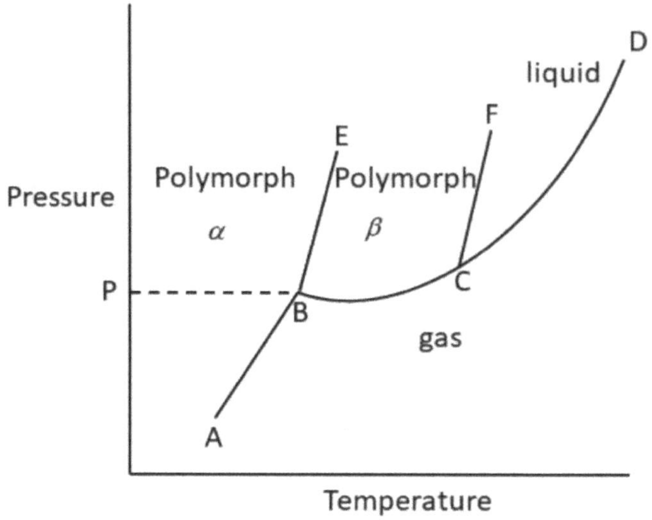

FIGURE 10.4 Pressure versus temperature in the phase diagram for a one-component system.

AB and BC are sublimation curves for α and β polymorph, respectively. The area enclosed by the curve BE is for polymorph α, and the area EBCF is for polymorph β, whereas area DCF represents the liquid phase. When the crystal of α polymorph is heated, it sublimes at a pressure below P and may transform to polymorph β above pressure P. Thus, two invariant points B and C are called triple points for which F = 0 and P = 3. It can be seen from the phase rule:

$$F = C - P + 2, \quad F = 1 - 3 + 2 = 0$$

The area enclosed within the points A, B, C, and D is the region for the gaseous phase.

10.6.2 TWO-COMPONENT SYSTEM

The minimum number of phases in any system is one. So from the phase rule, the maximum degree of freedom in a two-component system is three.

$$F = C - P + 2, \quad F = 2 - 1 + 2 = 3$$

10.7 SIMPLE EUTECTIC SYSTEMS

10.7.1 LEAD–SILVER SYSTEM

Lead and silver mix together in the liquid state. This can be illustrated by the phase rule and diagram. Melting point of pure lead is 327°C while that of silver is 961°C. Both are immiscible in the liquid state. If silver is added to lead, depression in freezing point takes place. Solid lead and solution of silver in liquid lead attain equilibrium.

The curve AC is the freezing point or melting point curve of lead. Along AC, there are two phases, that is, solid Pb and its solution with silver. Here, B is the melting point of silver. When lead is added to silver, freezing point of silver decreases along BC. The curve BC is the freezing or melting point curve of silver. C is the eutectic point at which the curve AC and BC meets. This point is nonvariant where the degree of freedom is zero. Experimentally, it has been found that the compositions at eutectic point C are as follows:

Ag—2.6%
Pb—97.4%
Temperature—303°C

In the region, ACB enclosed by the curves AC and BC one homogeneous liquid phase of Ag and Pb exists.

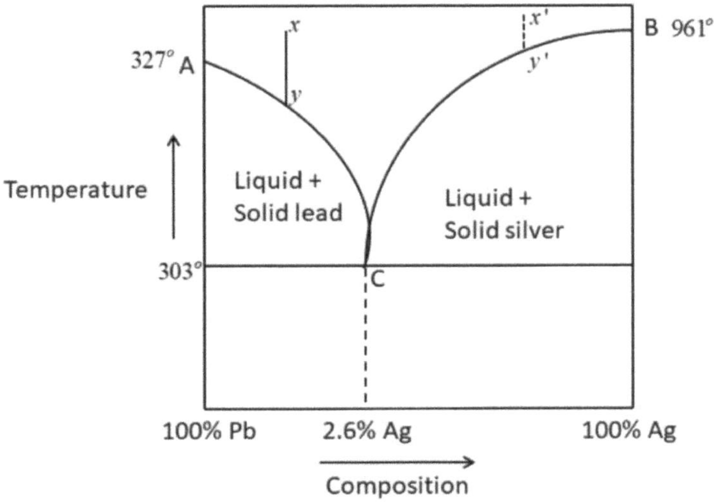

FIGURE 10.5 Phase diagram of lead–silver system.

When a liquid mixture of composition (x) is cooled, the temperature falls without any change in the composition until (y) on curve AC. At point (y) on the curve, AC lead solid is separated and the system becomes univariant. The temperature how decreases along AC with a change in composition. Lead solid separates and the solution becomes silver-rich. On cooling along AC at point C, the second solid starts crystallizing. The two solids lead and silver separates in a fixed ratio on further cooling. In the solid phase, the system has solid Ag and Pb both. Further cooling lowers the temperature below *DD'*. The region below the line *DD'*, two solid phases of lead and silver both exist.

Cooling along *x' y'* can be explained in the similar manner. At *y'* solid Ag crystallizes and the system becomes univariant. Further cooling along *y'c*, Ag separates and the solution becomes richer in Pb. At point C, three phases are present. Composition as well as temperature remains constant at this point.

10.8 BINARY SYSTEM WITH COMPOUNDS

Zinc and magnesium combine to form $Mg(Zn)_2$ in which Mg is 79% by weight and Zn is 21% by weight. Thus, compound melts without any change in the composition. The melting points of zinc and magnesium are 420°C and 650°C, respectively, and the congruent melting point of the compound is 590°C. The depression in freezing point of zinc is shown along the curve AC where zinc and liquid coexists. When magnesium is added to zinc, the behavior of phases is shown along AC. The curve DC shows the depression in freezing point of $Mg(Zn)_2$ by the addition of Zn. The curve DE shows the depression in freezing point on the addition of Mg. The curve BE shows the lowering of freezing point of Mg on adding zinc to magnesium. Here there are two eutectic points one at C and the other at E which is nonvariant. All three phases, that is, one liquid phase and two solid phases are present at this point. Similarly, there are three phases at E, that is, one liquid phase and two solid phases. This point is also nonvariant.

Vertical lines passing through A and B represent solids. Horizontal lines passing through C and E also represent the solid-state.

10.8.1 COOLING EFFECT

When cooling is done along the line xy, the temperature falls gradually. When it touches point y, $Mg(Zn)_2$ separates out. When the three phases are present, temperature and composition remain unaltered.

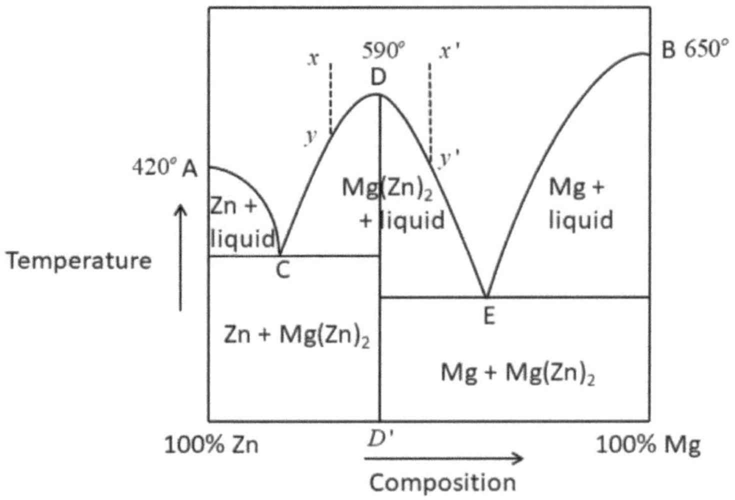

FIGURE 10.6 Phase diagram of Zn–Mg system.

In a similar fashion when a liquid of composition x' is cooled, the temperature drops along the line $x'y'$. At point y', the line $x'y'$ touches the curve DE and a new solid compound $Mg(Zn)_2$ separates.

The line DD' is represented on the phase diagram vertically which shows the temperature range over which compound $Mg(Zn)_2$ is stable. The compound $Mg(Zn)_2$ melts congruently because it changes from solid $Mg(Zn)_2$ to liquid of the same composition. The curves AC, CD, DE, and BE represented the liquids. The regions above this curve represent the liquid state.

10.9 BINARY SYSTEMS WITH SOLID SOLUTION

When the solids dissolve in one another to form a solid solution, we get completely miscible solid solution. The solid crystal obtained is homogeneous and is also known as an isomorphous mixture. This is why the solid compound has always definite composition.

Points A and B represent the melting points in the pure form of the metal, that is, copper and nickel. When B is added to A, the melting points of A are lowered or depressed. At low temperature, only one phase of solid solution exists, whereas a single phase of liquid solution exists at high temperature. At intermediate temperature, solid solution and liquid both exist. The compositions of the two phases are in equilibrium within this temperature range. The dotted line in the above-mentioned phase diagram touches the solids curve at point A which gives the composition of the solid solution. Some of the solid solutions are of anorthite, $CaAl_2Si_2O_8$, and albite $NaAlSi_3O_8$. The solid solution of both anorthite and albite are called plagioclase feldspars. Igneous rocks contain plagioclase feldspars. Thus, plagioclase crystals are rich in calcium at the center and sodium in the outer regions.

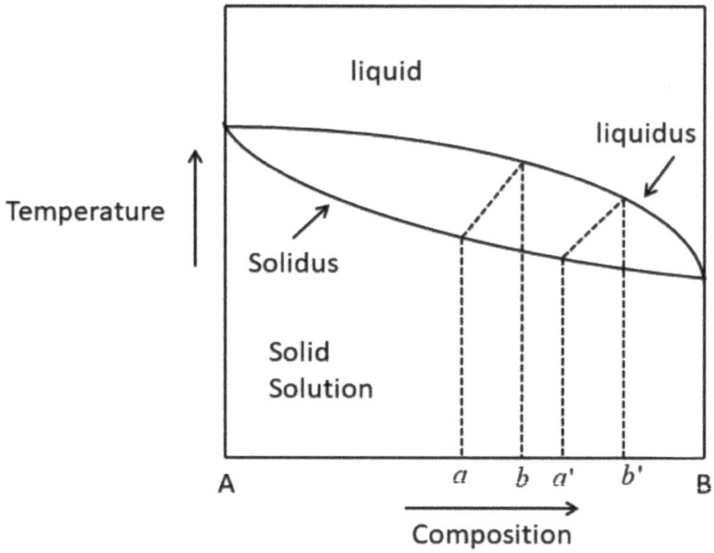

FIGURE 10.7 Phase diagram of completely miscible salt solution.

The phase diagram of a binary system with miscibility is applicable in copper–nickel solid solution. The melting points of pure copper and nickel are 1084°C and 1451°C, respectively. The addition of nickel to copper increases slowly the freezing points. The freezing point of all the mixture lies between 1084°C and 1451°C. Here the solids are continuous as melting points of the whole series of solid solutions lie between 1084°C and 1451°C. The region above the liquids is liquid and the region above the solids is solid. There is always a change of temperature during the process of solidification. Other systems having continuous solid solutions are Au–Pt, Co–Ni, Bi–Sb, Ag–Pd, Au–Ag, AgCl–NaCl.

10.10 BINARY SOLID SOLUTIONS WITH THERMAL MINIMUM

A solid solution may possess a lower melting point than other solid solutions of the series. They are completely soluble in both the solid and liquid states. But they have either a thermal minima or maxima in the liquids and solids curves illustrated in Figure 10.8. The thermal maxima and minima are known as indifferent points.

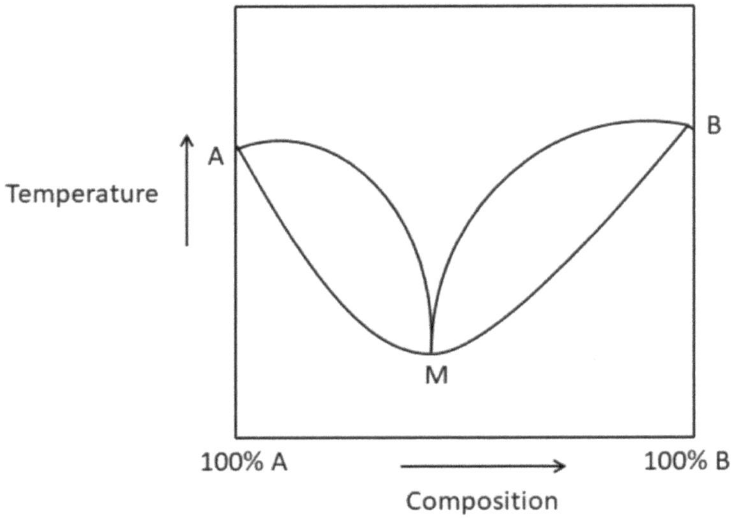

FIGURE 10.8 Binary solid solution with thermal minima.

Points A and B represent the melting points of two solid components. The two curves meet at a point M. At this melting point, liquid is in equilibrium with the solid of the same composition.

10.11 BINARY SOLID SOLUTION WITH THERMAL MAXIMA

In the case of thermal maxima also, the solidus curves meet the liquidus at the maximum point M. At this point, the liquid solution and the solid solution have the same composition. For this also, this liquidus and solidus are continuous through the thermal maximum. No discontinuity has been observed in such cases. The solid solubility occurs when the cations are similar in size, that is, Al^{3+} and Cr^{3+}. Thus for solids solutions, interatomic interactions should be the same. Similar types of atoms and similar sizes facilitate formation of solid solution.[7,8] At point M, the compositions of the components have a simple ratio, for example, 1:1, 1:2. With decreasing temperature, the extent of the solid solution is maximum at the solidus temperature M.

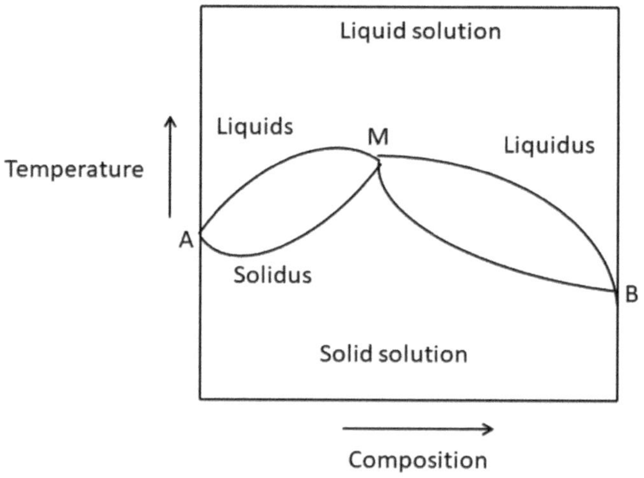

FIGURE 10.9 Schematic phase diagram with thermal maxima.

The region above the curve AMB represents the liquid solution. The curves AM and MB represent solidus and the region below AMB represents the solid solution. In the region, solid solutions of metals or components having comparable sizes exist. M is the thermal maxima in the phase diagram. At a fixed temperature at M, there should be three phases for equilibrium but in solid solution, there are never more than two phases, that is, solid solution and liquid.

10.12 BINARY SYSTEM WITH PARTIALLY MISCIBLE SOLID SOLUTION

10.12.1 EUTECTIC SYSTEM OF TWO SOLUTIONS

In this system, the freezing point of one component lowers due to the addition of the other component. The curves of lowering for both the component meet at point C where an eutectic of two solid solutions exists. At this point C, the liquid comes in equilibrium with two different solid solutions. In this system, the crystalline phases have partial solubility in each other crystal of one component B dissolve in crystal of A to form a solid solution. As the temperature decreases, the range of solid solution increases. This solid solution is maximum at point D. When the temperature is further lowered, the extent of solid solution decreases along the curve DF. Similarly, component A dissolve in crystal of B and maximum limit of this solid solution has been shown BEG. The maximum extent of this solid solution occurs at the point E.

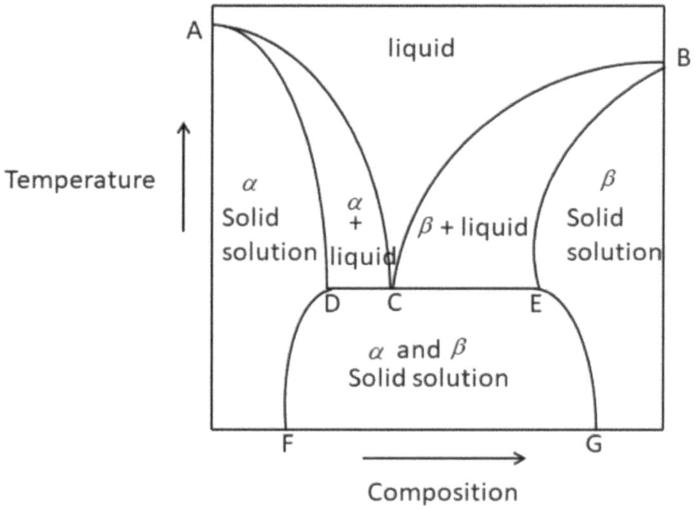

FIGURE 10.10 Binary eutectic system with partially miscible solid solutions.

The liquidus consists of the curves AC and BC. Curves AD, BE, ED, DF, and EG represent the solidus. Curves DF and EG show the change of the concentration of α-solid solution and β-solid solution with the lowering of temperature. At point C, a conglomerate of α and β is obtained. The concentration of α-solid solution is represented by D and β-solid solution by E. Thus, it is clear that two separate solid phases are in equilibrium with one liquid.

So from condensed phase rule equation, $F = C - P + 1$, $F = 2 - 3 + 1 = 0$

The composition of the solid changes along BE and the liquid along BC. At eutectic temperature, the solid and liquid have the composition

E and C, respectively. Some of the examples obeying this type of phase diagram are forsterite ($MgAl_2O_4$)–willemite (Zn_2SiO_4), spinel ($MgAl_2O_4$)–corundum (Al_2O_3), Cd–Sn, Fe–C, Au–Ni, Sn–Pb, AgCl–CuCl, etc.

10.12.1.1 BINARY SYSTEM WITH PARTIAL SOLID SOLUTION FORMATION

In this type of binary system, partial solid solubility has been shown in a somewhat different diagram.

An immiscibility dome exists within the solid solutions of α and β. It has an upper consolute temperature as shown in Figure 10.11. Above this upper consolute temperature, a single phase solid solution exists and below it, a mixture of two phases exists.

In another case, the dome expands to higher temperature and it intersects the melting curves.

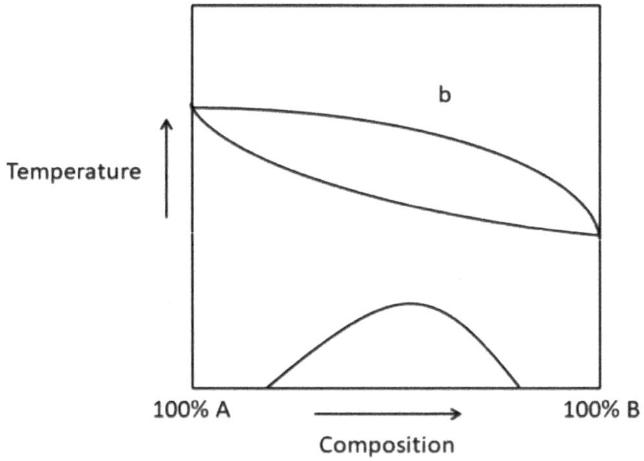

FIGURE 10.11 Binary system with partial solid solution formation.

10.13 BINARY SYSTEMS WITH SOLID–SOLID PHASE TRANSITION

Phase transitions are first order with thermodynamic view and a change in volume, enthalpy, and crystallography. Many primary bonds are formed and broken. Pure A or pure B, the two solid polymorphs coexists in equilibrium at a fixed points. The low-temperature polymorphs of *A* and *B* are represented as αA and αB, respectively.

The transition temperature for $\alpha B \Leftrightarrow \alpha B$ is T_1 and for $\alpha A \Leftrightarrow \alpha A$ it is T_2. Several examples of solid–solid phase transitions occur in the CaO–SiO_2 phase diagram. A number of bis(*n*-alkyl ammonium) tetra chloro metalloids (II) have been synthesized and solid–solid phase transitions have been studied using DSC and IR. Transition temperatures of C_nMn, C_nCo, C_nZn, and C_nCu lie between 28 and 86 KJ mol^{-1} and are used as potential thermal-energy storage materials.[9] The formula is represented as $(n\text{-}C_nH_{2n+1}NH_3)_2\ MCl_4$ (C_nM, n=10, 12, 16 and M= Cu, Zn, Hg, Mn, Co, Ni).

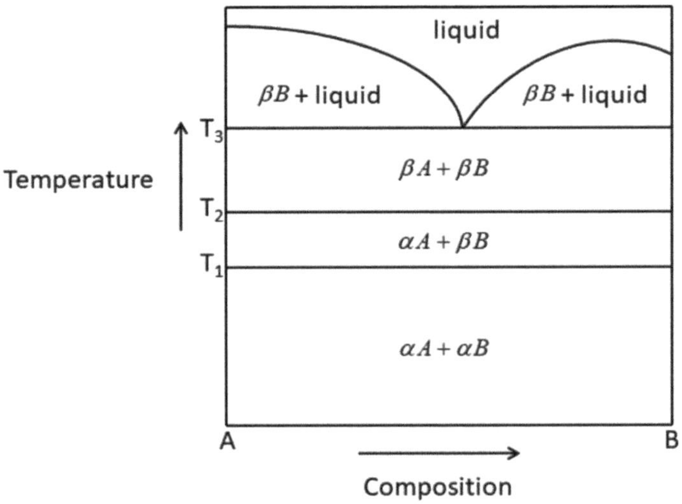

FIGURE 10.12 Simple binary system with polymorphic transitions.

10.14 PERITECTIC SYSTEM OF TWO SOLID SOLUTION

The system is transition reaction that involves lattice or morphotropic changes. It does not involve the formation or rupture of chemical bonds. The transformation of a solid solution takes place into another solution along with liquid. Interaction of one solid solution and liquid may take place to form another solid solution. The change can be represented as:

$$\alpha \rightleftharpoons \beta + \text{liquid}$$

It is clear from the diagram that the addition of component B to A raises the freezing point of A and addition of A to B lowers the freezing point. AC and AD are the curves that represent the liquidus and solidus of the α-phase. Similarly, BC and BE represent the β-phase. At a temperature t, the liquid composition at C is in equilibrium with two solid solutions of compositions D and E. So this temperature remains constant until the reaction is completed and this is also an invariant point known as peritectic. The reaction at this peritectic point is $\alpha \rightleftharpoons \beta$ + liquid and according to condensed phase rule F=0, that is, the degree of freedom is zero.

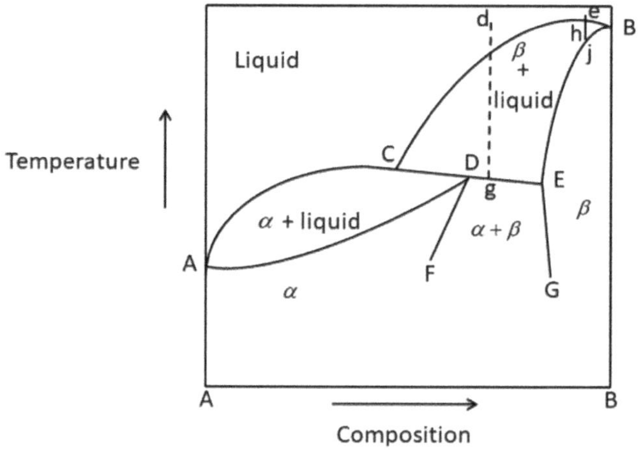

FIGURE 10.13 Phase diagram of peritectic system of two solid solutions.

The region above the curve ACB represents liquid . When a liquid is cooled from the point e, B solid solution separates at h on the curve BC representing liquidus BC. At the point d the system is completely solid. If the liquid is cooled from the point d, β-phase is deposited. When it is cooled further, the liquid undergoes composition change along the curve BC. The β crystals are represented along the solidus BE when it touches the line at g, Eg parts of β crystals are in contact with gc parts of liquid. The system undergoes a peritectic reaction at a constant temperature β + liquid $\rightleftharpoons \alpha$.

Again on cooling the composition of α and β-phase changes along DF and EG, respectively. The region DFGF constitutes a heterogeneous mixture of α and β-phases. When the temperature drops below the eutectic point, solid α phase is deposited. The curve AD shows the solidus. Some of the examples worth mentioning are LiCl–AgCl, $AgNO_3$–$NaNO_3$, steel, brass, etc.

Peritectics point is also very useful in the explanation of phase diagram of alloys. The equilibrium solid solubility of indium in Mg is around 19.4% at peritectic[10,11] temperature of 484°C. It decreases to 18.62% at 327°C and 13.95% at 200°C. The nature of the curves and peritectic points determine and explain the formation of alloys and their hardening.

10.15 FORMATION OF STEEL ALLOY

Iron shows polymorphism. At the same time, iron shows different behavior in different forms. The polymorphic forms of iron are given as:

i) Body-centered cubic α.
ii) Face-centered cubic γ.
iii) Body-centered cubic δ.

α form is stable below 910°C γ form is stable between 910°C and 1400°C and δ form is stable between 1400°C and 1534°C. γ polymorph of iron can dissolve around 2% by weight in solid solution formation. α and δ forms dissolve carbon from 0.02 to 0.1 weight%. Face-centered cubic γ structure is densely packed but the interstitial holes are suitable for occupation by carbon octahedral sites are available for carbon occupation. Octahedral sites are at cubic face centers in α-Fe and at the body center in γ-Fe and these interstitial sites are occupied by carbon. Most of the carbon steels contain even less than 1% carbon. When it is cooled from the melt in the temperature range 800–1400°C, a solid solution of carbon in γ-Fe is formed known as austenite.

When the steel is cooled quickly martensite forms are obtained instead of ferrite and cementite. Martensite is nothing but deformed austenite structure in which carbon atoms remain in the solid solution. The hardness of steel depends on cooling conditions. Martensite steels are very hard. If the steel is cooled gradually below 723°C, the decomposition is slow, and as a result pearlite texture is obtained. Thus, steel is an alloy of iron with low amounts of carbon, carbon steel, tool steel, stainless steel, and alloy steel are some of the common alloys of iron. Alloy steels contain other elements also such as vanadium, silicon, manganese, and chromium. The percentage of alloying elements provides the varying mechanical properties. The alloying can be understood from the Fe–C phase diagram. This diagram explains ferrite formation, cementite formation, and martensite formation. Cementite, Fe_3C, is thermodynamically metastable but kinetically stable.

10.16 THREE COMPONENT CONDENSED SYSTEM

In a three-component or ternary systems, pressure, temperature, composition of one component, and the composition of other components are the four independent variables. At a certain fixed temperature, three crystalline phases and liquid coexist. Ternary systems are represented by equilateral triangles whose corners are represented by three components. Temperature is represented by the vertical axis. A three-dimensional prism displays the effects of varying temperature and composition. Each of the three binary edges is divided into a hundred divisions. Each apex of the triangles is taken as 100% of the component. Apex A represents 100% of A, B represents 100% of B, and C represents 100% C. Sides AB and AC are divided into 100 equal parts. The lines dividing the sides BA and BC and parallel to AC gives the percentage of B. The lines CA and CB, parallel to AB represent the percentage of C. Compositions can be in weight percent, mole percent, and atom percent. Generally, mole percent is preferred for inorganic systems. There are three apexes. Any point on any side of the triangle gives rise to a binary system. The point within the triangle represents a ternary system of compositions A, B, and C. For example, a ternary system may contain 30% A, 20% B, and 50% C, making a total of 100%. In a similar manner, other points within the triangle may contain different percentages of A, B, and C.

This method is also applied to determine phase composition in the triangle that is not equilateral.

10.17 SIMPLE EUTECTIC SYSTEM IN TERNARY COMPOUNDS

The equilibrium diagram for a ternary system containing solid and liquid phases has been shown by a right prism. The melting points of the three components have been shown by A, B, and C. The point e_1 is the eutectic point for the component A and B, e_2 is the eutectic point for the components A and C, and e_3 is the eutectic point for C and B.

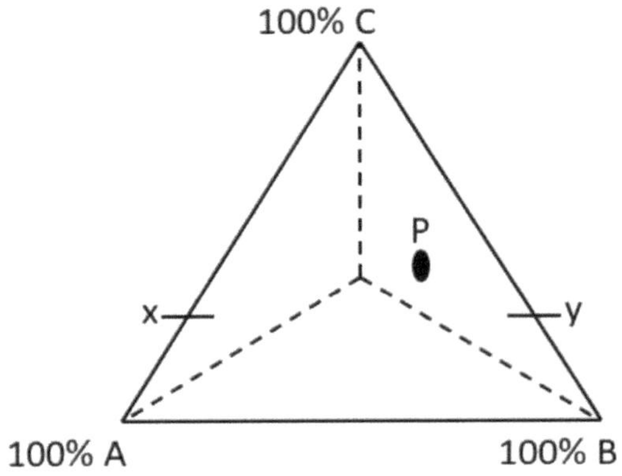

FIGURE 10.14 Equilateral triangular.

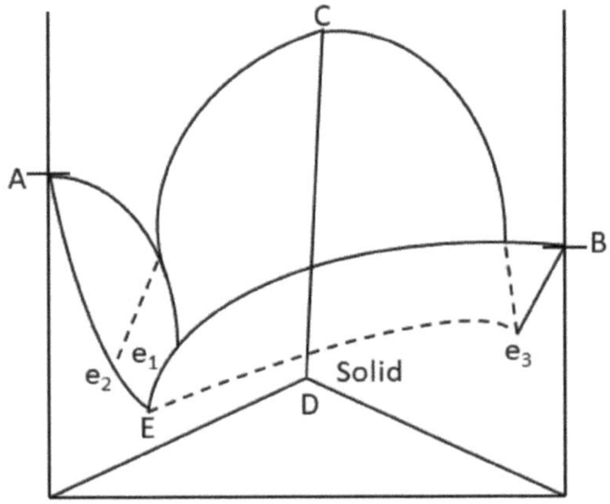

FIGURE 10.15 Eutectic point for ternary system.

When C is added to the eutectic mixture of A and B at e_1, the temperature is lowered along e_1E. At point e_1, two solid phases A and B coexist in equilibrium with the liquid. When B is added to the eutectic mixture of A and C at e_2, the temperature is lowered along e_2E, e_3E represents the equilibrium between B and C and ternary solutions.

The point E is known as the ternary eutectic point where three solid phases A, B, and C exist and are in equilibrium.

The pure solids and ternary liquids exist together below the liquidus surfaces e_1E, e_2E, and e_3E. The whole system consisting all three components is solid below the eutectic point. The temperature of each point has been specified as the diagram is not isothermal. Pb, Bi, and Sn system, a ternary system, is explained by this diagram.

10.18 TERNARY SYSTEMS CONTAINING BINARY COMPOUNDS

In this system, one congruently melting binary compound exists. The triangle ABC can be divided into two triangles A-B-BC and A-BC-C obtained by drawing a line perpendicular from vertex A to the base. Each triangle has been treated separately for the application of the phase diagram. The line A-BC is a true binary join. The composition of solid, liquid, or a mixture of solid and liquid can be represented along this line.

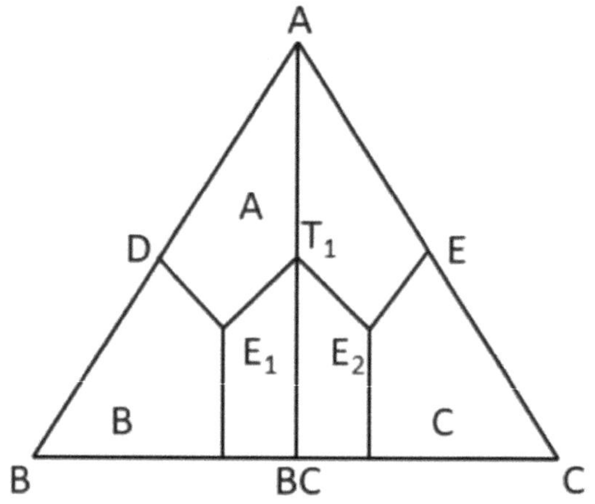

FIGURE 10.16(a) Melting relations in ternary systems containing binary phase.

The eutectic is here at T_1 which is the lowest temperature where the two lines join. It can also be said as thermal maximum. $T_1 > E_1$, E_2, and so this maxima separates the primary phase fields of A and BC in the ternary system A-B-C.

The region above DE_1T_1 and T_1E_2E represents A and the region below BCE_2E represents C. The region below DE_1 represents B. Geographically, it can be said that T_1 is the height of a line passing two mountains A and BC.

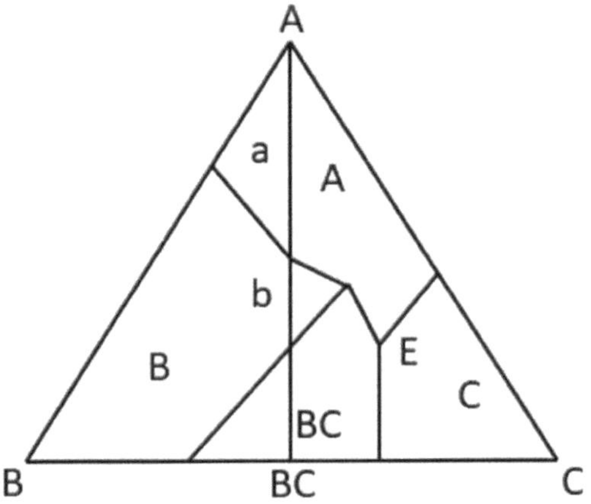

FIGURE 10.16(b)

In this figure, the univariant curves separate the phase fields of A and B and B and BC. These curves cross the line A-BC. So the invariant point is outside the triangle of AB-BC and inside the triangle A-C-BC. At this invariant point, all the three crystalline phases coexist. This invariant point lying outside its own triangle is known as ternary eutectic point. If the invariant point is inside its own triangle, it is a ternary eutectic point. The peritectic temperature P is always higher than the eutectic temperature E. Figure 10.16(c) can also explain the nature of join A-BC.

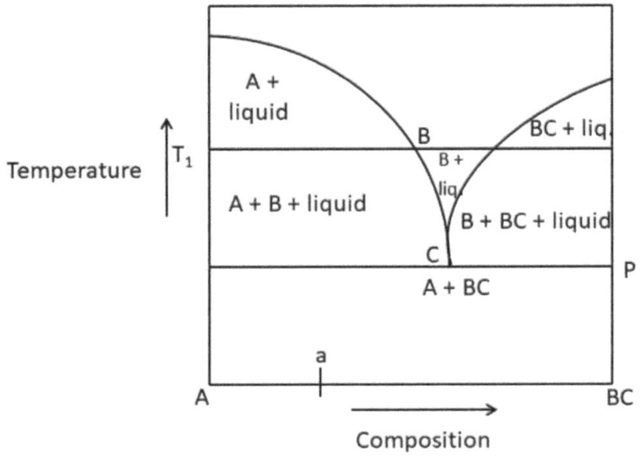

FIGURE 10.16(c)

When the liquid composition **a** is cooled, the composition is within the primary phase field of A and A crystalline phase appears first. A continues to crystallize at temperature T_1. By this time, the liquid composition arrives at **b** where a crystal of B starts separating out. The liquid composition departs from line A-BC and enters the region A + B + liquid. When the temperature is dropped further, A and B crystallize until the peritectic temperature P is reached. Below point P, the equilibrium subsolidus A + BC is present. From Figure 10.16b, it is clear that the crystallization occurring on cooling takes place up to P for the compositions within the triangle A-B-BC. Thus, P is said to be solidus temperature for the triangle A-B-BC. In the triangle A-BC-C, E is the eutectic temperature or the solidus temperature.

10.19 SUBSOLIDUS EQUILIBRIUM

The system containing many binary phases is at subsolidus temperatures and remains divided upto a number of smaller triangles. Suppose there are two binary compounds AC and BC and are present in the compatibility triangle ABC in a ternary system. The subsolidus triangle AC-BC-C exists. In the remaining diagram, there may be A and BC as compatible phases or B and AC. If A and BC are compatible, the line joining A and BC represents the stability in the presence of each other. There are two three-phase triangles A-B-BC and A-BC-AC. In both the triangles, all the ternary phases coexist. B and AC are incompatible phases and so mixtures of B and AC react to give a mixture of A and BC.

During the manufacture of Portland cement clinker, this phase diagram finds good application. Ca_2SiO_4 and Ca_3SiO_5 are the desired silicates for the manufacture of clinker. Additives are not added as it may react with either Ca_3SiO_5 or Ca_2SiO_4 to give other undesired silicates. Similar phase diagrams are applicable to refractory bricks and molten slag.

10.19.1 MECHANISM OF PHASE TRANSITION

Classification of phase transition has to be discussed before going into the details of phase transition. There are two types of transition, that is, reconstructive and displacive transitions. Reconstructive transition involves the reorganization of crystal structure caused by breaking and formation of new bonds. When graphite is changed to diamond, this involves a complete change in crystal structure. Bond lengths and bond angles change completely. Hexagonal sheets are present in graphite and tetrahedral four coordinates carbon atoms are in diamond. So there is a complete change of structure during the phase transition of graphite to diamond and as a result the transition is reconstructive.

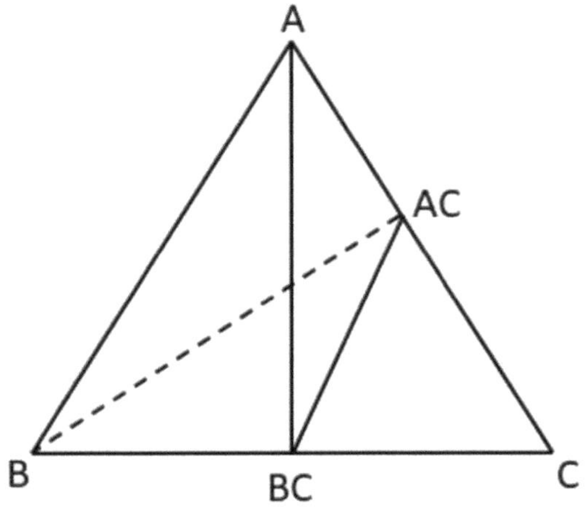

FIGURE 10.17 Compatibility triangles in a ternary system at subsolidus temperature.

One more cited example is quartz to cristobalite transition in SiO_2. The polymorphs have the different three-dimensional framework. High activation energy is required for reconstructive transition and as a result the phase transition takes place very slowly.

Displacive phase transitions involve the distortion of bond instead of the formation of new bonds. Obviously, small activation energy is required and as a result the transition becomes faster.

Silica: Quartz also undergoes displacive phase transitions. There is a symmetry relationship between the two polymorphs.

10.20 ORDER–DISORDER TRANSITIONS

This type of transition involves exchange of atoms or ions. Order and disordered polymorphs of $LiFeO_2$ are stable below and above 700°C. The disordered polymorph may be preserved to room temperature. It has a rock salt structure in which octahedral sites are occupied by Li^+ and Fe^{+3} ions. When the disordered $LiFeO_2$ is heated at 600°C, the order of oxide ion remains unchanged and the cations order over the octahedral sites. During ordering migration of cations takes place. However, ordering transition in $LiFeO_2$ is slow. Li_2TiO_3 also exhibits order–disorder transition. The equilibrium transition temperature is around 1200°C. Rapid transition takes place below 1200°C.

One of the phase transition type is bond type in which grey tin is converted to white tin. Semiconductor changes to the metallic character. In the transformation of diamond to graphite, insulating behavior changes to semiconducting behavior of graphite. Such types of changes have been thus observed in carbon and tin. The important mechanisms of phase transition have been described as (1) nucleation and growth transitions, (2) order–disorder transitions; (III) martensitic transitions.

10.20.1 NUCLEATION AND GROWTH TRANSITIONS

In nucleation, nuclei of the new phase possessing a critical size aggregate together in the parent phase known as nucleation. This is known as nucleation-growth transitions.

Order–disorder transitions have been explained above which involve positional and orientational disordering as well as a disordering of electronic or nuclear spin states.

A martensitic transition is caused by the displacement of atoms which corresponds to a homogeneous deformation. Martensitic transitions occur very fast. In CsCl–NaCl, a martensitic transition has also been observed between the two phases. Though martensitic transitions have been observed in steel, several inorganic solids undergo this type of transition. Some of the examples worth mentioning are ZrO_2 and V_3Si. All the phase changes do not undergo any change in composition during phase transition.

10.20.2 KINETICS OF PHASE TRANSITION

The thermodynamic parameters, for example, Gibbs free energy and entropy have been discussed in the beginning of the chapter. But the thermodynamic property changes do not give information regarding the rate of reaction. There are some transitions that take place in both the forward and backward directions without any hysteresis. Some transitions take place with some hysteresis. High-temperature polymorph is transformed to low-temperature polymorph. Rates are slow in either direction close to the equilibrium transition temperature T_c. Gibbs free energy change is little driving force for the transition to occur either in the forward direction or backward direction. At some temperature Tm, the high-temperature form II is converted to form I on cooling. On further cooling, the rate of phase transition from II to I again decreases. The rate of transition from II to I increases upto the region C_1. But there is no corresponding maxima above Tc in the I → II transition.

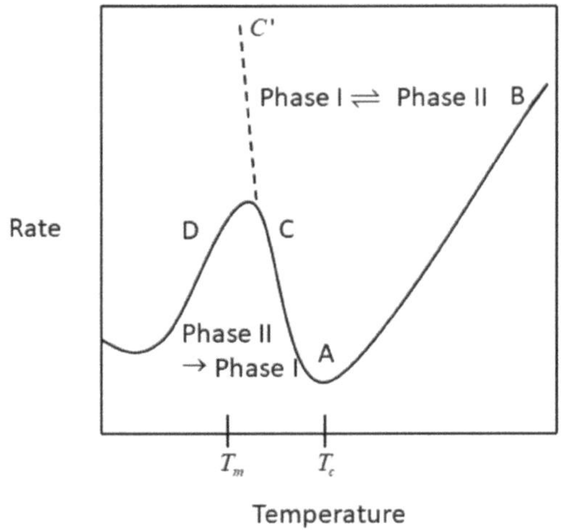

FIGURE 10.18 Variation of rate with temperature for phase transition.

Arrhenius equation can here be applied.

$$k = A.e^{-\frac{E_a}{RT}}$$

$$\ln k = \ln A - \frac{E_a}{RT}$$

where k is the rate, E is the activation energy, T is the absolute temperature, and R is the gas constant.

If ln*k* is plotted against $\left(\frac{1}{T}\right)$, the slope of the line gives the value $\left(\frac{-E_a}{R}\right)$ and intercept is equal to ln*k*. For a phase transition to occur, the minimum amount of energy has to be acquired, that is, activation energy barrier has to be overcome by phase I or II.

The other important factor is the difference in Gibbs free energy change for the two phases.

$$\Delta G_{I\text{–}II} = G_I - G_{II}$$

Where G_I and G_{II} are the Gibbs free energy of the polymorphs I and II, respectively. So the Gibbs free energy change is a measure of the spontaneity of the phase transition. If ΔG is negative, the phase transition is preferentially observed. But at $\Delta G = 0$, the phase transition cannot take place in any direction.

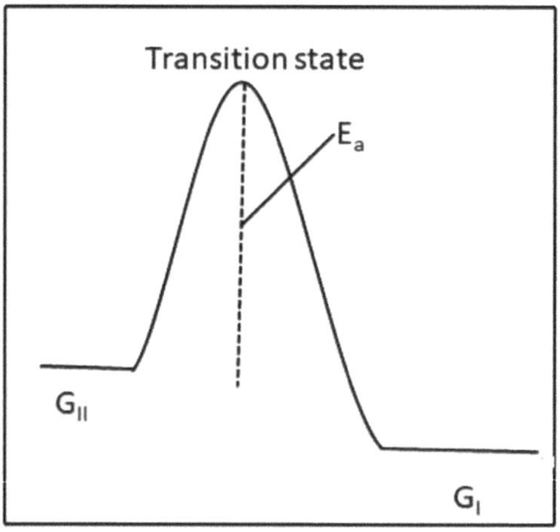

In the case of ideal transition, enthalpy and entropy of polymorphs are independent of temperature.

$G = H - TS$ (from thermodynamics)

$$\Delta G = \Delta H - T\Delta S \quad (10.1),$$

where ΔH is the enthalpy change, and ΔS is the entropy change.

$$\Delta G = (T_c - T)\,\Delta S \quad (10.2)$$

The magnitude of ΔG at temperature T is a function of the difference in temperature $T_c - T$)

$$\Delta G = \frac{(T_c - T)}{T}\Delta H$$

Many phase transitions have been supposed to take place by the mechanism of nucleation and growth. Nucleation of the phase takes place slowly. But there is no suitable quantitative explanation for certain parameters.

10.20.3 CRITICAL SIZE OF NUCLEI

The nuclei of the product phase are formed initially in the nucleation and growth transition. The free energy change accompanying phase transition is here due to the formation of the product phase in the nuclei and due to the free energy of the nuclei.

The change in free energy ΔG is given by

$$\Delta G = -\frac{4}{3}\pi r^3 \Delta G_v + 4\pi r^2 y \tag{10.3}$$

The first term on the right-hand side shows the decrease in bulk free energy, whereas the second term shows the surface free energy. If radius r of the spherical nucleus is small, ΔG will be positive because $4\pi r^2 y$ is greater than $\frac{4}{3}\pi r^3 \Delta G_v$, and as a result the nucleation becomes thermodynamically unstable. Thus for being stable, the nucleus should have a certain minimum size, for example, critical size. When the nucleus attains a critical size, ΔG starts decreasing and the process of nucleation becomes favorable. Once the nuclei size r is greater than r_c (critical size), they start growing.

The critical radius can be achieved by differentiating ΔG with respect to r and making it equal to zero.

$$\frac{d\Delta G}{dr} = 0$$

$$\frac{d\Delta G}{dr} = 8\pi r y - 4\pi r^2 \Delta G_v \tag{10.4}$$

$$\text{or, } 8\pi r y = 4\pi r^2 \Delta G_v \tag{10.5}$$

$$\text{or, } \Delta G_v = \frac{8\pi r y}{4\pi r^2} = \frac{2y}{r} \tag{10.6}$$

$$\text{or, } r_c = \frac{2y}{\Delta G_v} \tag{10.7}$$

Putting the value of r_c into Equation 10.3, we get

$$\Delta G = \frac{16}{3} \frac{\pi y^3}{\Delta G_v^{\ 2}}$$

Nucleation is difficult around the temperature T_c. Thus, ΔG is positive at the thermodynamic transition temperature and as a result nucleation is difficult at T_c. Thus in order to have a transition from one phase to another, there must be a little supercooling or superheating. The Gibbs free energy change nucleation has been shown to vary with the radius *r* in Figure 10.19.

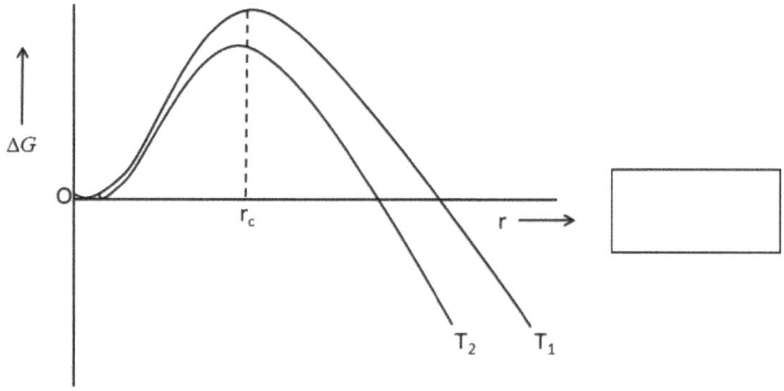

FIGURE 10.19 Gibbs free energy change as a function of radius.

After the completion of nucleation, phase transition takes place by the growth of the nuclei having critical size (r_c).

The nucleation rate is given by the basic Arrhenius equation:

$$k = A.e^{-\frac{E_a}{RT}}$$

where E_a is the overall activation energy which is here a combination of ΔG_c (critical free energy), and G_a is the activation energy for the jump of individual atoms during nucleation

$$k = A.e^{\left[-\frac{(\Delta G_c + \Delta G_a)}{RT}\right]}$$

The nucleation also depends on the impurities, defects, dislocations, and surfaces. When all the parts of the parent phase are identical, the nucleation is homogeneous and random in nature. Heterogeneous nucleation takes place preferentially at the defect centers.

10.21 AVRAMI EQUATION

Avrami established an empirical relation for nucleation-growth transition. Avrami equation can be expressed as

$$\alpha = 1 - e^{(-kt)^n} \tag{10.8}$$

where α is the volume fraction of the product or transformed phase, k is the rate constant, and n varies from 1 to 5 depending on the nature of the nucleation and growth.

Taking log twice of Equation 10.8, we have the following equation:

$$\log.\log\left(\frac{1}{1-\alpha}\right) = n\log k + n\log t + \log.\log e$$

When $\log.\log\left(\frac{1}{1-\alpha}\right)$ is plotted against log*t*, straight line is obtained.

The utility of Avrami equation lies in the fact that the mechanism of transformation of phases can be explained to some extent. The values of n for polymorphic transition give an insight into the mechanism of nucleation. It has been found that nucleation occurs at the beginning of phase transformation for values of $n = 3$. If $n = 4$, the nucleation continues to occur in untransformed phase. The transition of grey tin to white tin may be explained by nucleation-growth mechanism having the n value of Avrami equation between 1.5 and 2. For phase transition of white to grey tin, the value of Avrami n is equal to 3. Avrami equation also explains order–disorder transition. When the transformation is near completion, very little untransformed material becomes available for fresh nucleation.

The characteristics of Avrami equation are as follows:

i) Nucleation takes place over the entire phase of the solid at a uniform rate.
ii) Nucleation and growth take place in all the orientation or directions uniformly.

In the beginning, nucleation is random and the growth goes on increasing which leads to a high value of Avrami n equal to 4. When the sites for nucleation are occupied, formation of new particles stops immediately. The values of n may be equal to 1, 2, 3, and 4.

The value of Avrami constant (n) can be obtained from the graph, as given in Figure 10.20.

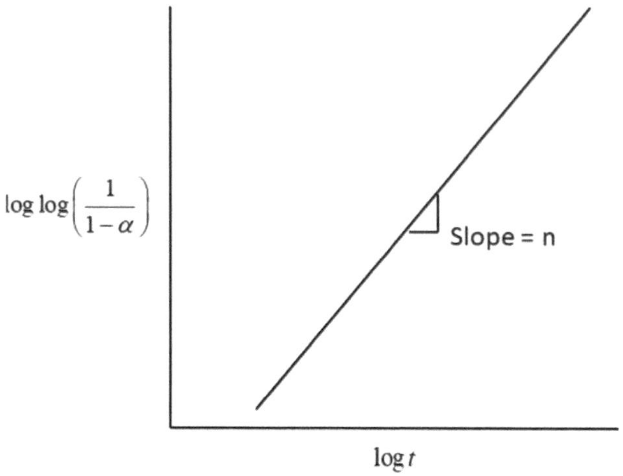

FIGURE 10.20 Avrami plot.

Thus, Avrami equation and plot explains successfully the phase transformation at a constant temperature. It explains the kinetic aspect of crystallization. Thermodynamic criteria for spontaneity of phase transformation are the same as applicable to general reactions.

10.22 FACTORS AFFECTING KINETICS OF PHASE TRANSITION

During the phase transition, a large number of atoms are involved to form the stable nucleus. The anatase form of TiO_2 changes to rutile form through nucleation-growth transition. In this transformation, nuclei of rutile is formed at the surface and the bulk of the anatase crystals. With the increase of time, growth of nuclei takes place:

i) The composition of the phases remains unaltered during a phase transition. The surface area of the sample and temperature of the system are important factors.
ii) The surface energy of a nucleus and total surface area of the original crystals. Powdered sample provides more surface area.
iii) Equilibrium transition temperature
iv) As the bonds are formed and broken, activation energy is the key factor that affects phase transition.

v) Phase transformation is accompanied by a change in volume. So change in volume along with pressure is important factor that affect the kinetics of phase transition.
vi) The Arrhenius factor A used in the Arrhenius equation also affects the kinetics.
vii) Impact of polymers on phase transition kinetics[12].

From transition state theory, the difference in volume between two polymorphs can be related to rate by the equation given below:

$$\log k = C - \frac{\Delta V.P}{RT}$$

where ΔV is the difference in the volume of the initial polymorph and the transition state. R is the universal gas constant, T is temperature, k is rate, P is pressure, and C is constant.

10.23 ENTROPY CHANGE IN ORDER–DISORDER TRANSITIONS

It has been established experimentally that order–disorder phase transitions are thermodynamically first order and discontinuous. The transition may be of second order and continuous in case of disordering over a large temperature range. The magnitude of entropy change is related to the structural changes. If the structures of ordered and disordered polymorphs are known, the configurational entropy may be calculated. From experimental data, silver iodide transforms from wurtzite to body-centered cubic followed by entropy increase. The entropy change at the transition is given by:

$$\Delta S = R \ln \frac{n_2}{n_1}$$

where ΔS is the entropy change, R is the universal gas constant, n_2 and n_1 are the number of configurations in the two polymorphs.

In silver iodide (AgI), an Ag^+ ion may be placed in any one of the 12 positions and so $n_2 = 12$.

But in β polymorph of AgI, only two positions are available for occupation and so $n_1 = 2$

$$\Delta S = R \ln \frac{12}{2} = R \ln 6 = R 2.303 \log 6$$

Putting the values of R and log 6 into the above equation, ΔS becomes equal to 14.7 $J.mol^{-1}K^{-1}$.

The effect of the maximal entropy change can be derived from the magnetization on measurements. Magnetic entropy change has also been observed in alloys. Thus, entropy change is associated with martensic–austensic structure transition.[13–16]

Thus, phases with more structural order show less entropy, whereas fluids show greater entropy under similar thermodynamic conditions. Ordered and disordered phases may be assessed with the relative amount of entropy.

KEYWORDS

- **phase transition**
- **Gibbs free energy**
- **one component system**
- **binary system**
- **critical size**
- **Avrami plot**

REFERENCES

1. Kaldis, E. Growth of Mercuric Iodide Bulk Crystal. In *Encyclopedia of Materials: Science and Technology*; 2001.
2. Sebastian, M. T. Low Temperature Cofired Ceramics. In *Dielectric Materials for Wireless Communication*; 2008.
3. McGreevy, R. L. Reverse Montecarlo Methods for Structural Modeling. In *Computer Modeling in Inorganic Crystallography*; 1997.
4. Nagara, H. Structural Phase Transitions in Dense Hydrogen. In *Strongly Coupled Plasma Physics*; 1990.
5. Bragas, A. V.; Merlin, R. Quantum Optics|Squeezed Phonons in Solids. In *Encyclopedia of Modern Optics*; 2005.
6. Olander, D. R. *General Thermodynamics*; CRC: Boca Raton, 2008.
7. White, M. A. Phases and Phase Equilibrium. In *Encyclopedia of Condensed Matter Physics*; 2005.

8. Burton, B. P. Order-Disorder Phenomena and Phase Separation. In *Encyclopedia of Materials: Science and Technology*; 2001.
9. Li, W.; Zhang, D.; Zhang, T.; Wang, T.; Ruan, D.; Xing, D.; Li, H. Study of Solid-Solid Phase Change of $(n\text{-}C_nH_{2n+1}NH_3)_2MCl_4$ for Thermal Energy Storage. *Thermochem. Acta* **1999,** *326* (1–2), 183–189.
10. Nie, J-F. *Physical Metallurgy*, 5th ed., 2014; 20.4.1.8.1 Mg-In-Ca Alloys.
11. Boettinger, W. J.; Banerjee, D. K. Physical Metallurgy, 5th ed.; 2014, 7.8.3.3 Rapid Solidification of Peritectic Systems.
12. Raina, S. A.; Alonzo, D. E.; Zhang, G. Z.; Gao, Y.; Taylon, L. S. Impact of Polymers on the Crystallization and Phase Transition Kinetics of Amorphous Nifedipine During Dissolution in Aqueous Media. *Mol. Pharma.* **2014,** *11* (10), 3565–3576.
13. Onuki, A. *Phase Transitions Dynamics*; Cambridge University Press, 2002.
14. Anderson, V. J. HNW Lekkerkerker—Insight Into Phase Transition Kinetics from Colloid Science. *Nature*, 2002. nature.com
15. Hu, F-X.; Shen, B-G.; Sun, J-R. Magnetic Entropy Change in $Ni_{51.5}Mn_{22.7}Ga_{25.8}$ alloy. *Appl. Phys. Lett.* **2000,** *76* (23), 3460–3462.
16. Guo, Z. B.; Zhang, J. R.; Huang, H.; Ding, W. P.; Du, Y. W. Large Magnetic Entropy Change in $La_{0.75}Ca_{0.25}MnO_3$. *Appl. Phys. Lett.* **1997,** *70* (7), 904–905.

CHAPTER 11

Solid Surface, Heterogeneous Catalysis, and Adsorption

ABSTRACT

Solid surfaces act as heterogeneous catalysts for a variety of reactions. A number of adsorption isotherms such as Freundlich, Langmuir, Temkin, Brauner, Emmet, and Teller have been proposed to explain the behavior of adsorption on the bentonite, montmorillonite surface. The derivations of these isotherms have also been done in this chapter to have a better understanding of the topic. In addition, Lagergren model and Dubinin–Radu Shkevich isotherms have also been incorporated. Origin and occurrence of bentonite, zeolites and their composition have been explained.

11.1 INTRODUCTION

Solid surface is the region where the bulk phases end at the interface. An atom or molecule at the surface has residual force. The field of forces of the atom at the crystal surface extends in space. The atom or molecule gets attached to the surface known as adsorption. In addition to adsorption properties, surface of solids also acts as catalyst in the case of heterogeneous catalysis. A number of solids are being used as catalyst in chemical industries or in laboratories for advanced research. Some worth mentioning catalysts are metal, for example, Ni, Pd, Pt (d-block element), alloys, clays, and clay minerals, for example, kaolinite, bentonite, montmorillonite, nitrides, and carbides. The physical states of such catalysts are essentially solids either crystalline or amorphous. The activity of the catalyst is interpreted as affecting the chemical reaction with a small amount of catalyst

Solid-State Chemistry: A Modern Approach. Ashok Kumar Jha, PhD (Auth.)

and achieving the equilibrium in a short time. Catalysts remain unaltered at the end of the reaction. It has been established that transition metal or d-block elements show catalytic properties and as a result, these elements are widely used as catalysts due to different oxidation states and vacant d orbitals. Owing to the availability of vacant d orbitals, more surface area is provided for reaction. Catalysts are undoubtedly selective producing certain types of products. These are directed to produce particular reaction products. It has been experimentally observed that different catalysts with the same reactants give different products. It is probable that one catalyst inhibits a particular reaction but accelerates other reaction. One example may be cited as H_2 and CO giving methane with nickel. But H_2 and CO give methanol with zinc oxide and chromium oxide as catalyst. Propylene and oxygen react over bismuth molybdate to give acrolein but when bismuth oxide is used as a catalyst on the same reactants, benzene, and hexadiene are obtained. Catalysts used in industries require regeneration. Zeolites and natural bentonites are used as catalysts in several reactions.

Uranyl antimonate is used for ammoxidation of propylene. Polymerization of ethylene and propylene takes place trialkyl aluminums and titanium (IV) halides popularly known as Ziegler–Natta catalyst.

In Haber's synthesis of ammonia, N_2 and H_2 are the reactants used in the presence of ferric and aluminum oxide as catalyst and potassium oxide as promoter.

$$N_2(g) + 3H_2(g) \rightleftharpoons 2\,NH_3$$

Thus, activity and selectivity of catalyst in promoting the reactions are being considered. Among d block elements, transition metal platinum is very widely used both in oxidizing and reducing conditions. Platinum is useful in dehydrocyclization, isomerization, and hydrogenation. Nickel also finds its use as a catalyst in hydrogenation of ethane.

11.2 ALLOYS

Alloys are also being used as catalysts with varying properties. The surface composition of alloys is different from bulk composition in certain cases. Alloys as a catalysts are also selective and specific for particular reactions. When nickel is alloyed with copper, its activity for hydrogenolysis is decreased. But for cyclohexane dehydrogenation the activity of Ni–Cu alloy catalyst remains unaltered. A heterogeneously catalyzed reaction involves (1) physisorption, (2) chemisorption of reactants on the surface, (3) diffusion at the surface, and (4) desorption and diffusion of products from the surface. The intermediates are formed at the surface which then change to the product.

11.3 PHYSISORPTION

The reacting gases collect on the surface of the solid catalyst to form a film. As the concentration is high, reaction takes place faster. Gaseous molecules are held by the surface by Van der Waals force. Physisorption is accompanied by low heat of adsorption.

11.4 CHEMISORPTION

In chemisorption, gaseous molecules are held to the surface by a valence bond. In this case, the surface compound is held on the surface of the solid. Chemisorption is accompanied by heat released. Chemisorption of hydrogen on nickel is shown as:

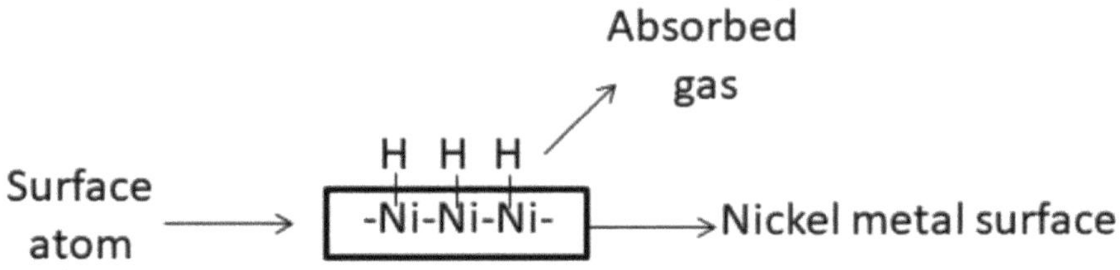

According to adsorption theory, two reacting gases A and B are first absorbed on the surface of the solid catalyst. The gases form activated surface complexes.

$$A + B + S \text{ (solid surface)} \rightarrow ABS \text{ (activated surface complex)}$$

$$ABS \rightarrow P + S$$

The activated complex is desorbed to give the product. The formation of SO_3 by the combination of SO_2 and O_2 in presence of a platinum catalyst takes place by this mechanism.

$$2SO_2 + O_2 \xrightarrow{Pt} 2SO_3$$

11.4.1 ACTIVE SITES

The catalytic activity of a solid is not uniform all over the surface. A number of edges, defects, cracks, and imperfections are present on the surface. Thus, active sites are associated with defects or irregularities. The atoms present at these sites have unshared electrons which make the chemisorption easier by formation of activated complex. In polymerization of propylene,[17–18] Ti^{3+} are the active centers in α-$TiCl_3$. In WS_2, W^{3+} ions act as active sites for hydrogenation of benzene.

11.4.2 ENERGY OF ACTIVATION

When catalysts are added to reactants, the energy of activation or energy barrier is lowered. With the lowering of the energy barrier, the rate of the reaction is increased.

11.4.3 PROMOTERS

It has been assumed that promoters enhance the activity of catalyst by creating irregularities on the crystal surface. The irregularities such as steps and kinks on planes of a single crystal of solid, for example, platinum increases reactivity. The role of steps and kinks on the surface of the catalyst in increasing the rate has also been confirmed by experiments.

11.5 POISONING OF A CATALYST

When the preferential adsorption of the unwanted substance takes place at the active centers, the reactants are hindered to reach those preoccupied centers and as a result, the activity of the catalyst is lowered. For example, acetanilide prevents the decomposition of hydrogen peroxide catalyzed by the walls of the containing glass vessel. Hydrocyanic acid prevents the decomposition of H_2O_2 by colloidal platinum. In the contact process of manufacture of concentrated sulfuric acid, SO_2 and O_2 give SO_3 in the presence of a platinum catalyst. But arsenious oxide slows down the formation of SO_3. Arsenious oxide here covers the surface of the solid, that is, active centers and as a result the catalyst is hindered. For transition metal, catalyst catalytic activity is related to percentage d-character of the metallic bond.[19–21] The second and third rows transition metals are also used as catalyst for industrial purposes. Dowden (1972) has provided a connection between the heterogeneous catalysis and coordination chemistry of transition metal compounds.[22]

Transition metal–MOS_2 catalysts supported on Al_2O_3 show the presence of an active transition-metal–MO–S' state. At these sites, Co or Ni form short bonds with sulfur.[23] In addition to transition metals of the second and third rows, several metal oxides, for example, V_2O_5 are used for oxidation reactions. CeO_2, Rh, Pd, and Pt are also used as auto-exhaust catalysts.

Pervovsite oxides of the type $CaMnO_3$–S and $La_{1-x}A_xMO_3$ (A = Ca, Sr; M = Co, Mn) are also used as catalysts.

$VOPO_4$ (Phosphorous vanadium oxide) catalysts are used for the oxidation of butane to maleic anhydride.[24]

Bismuth molybdates catalyze selecting oxidation of propylene to give acrolein.

$$CH_3CH{=}CH_2 + O_2 \xrightarrow[\text{molybdates}]{\text{Bismuth}} CH_2CHCHO + H_2O$$

11.6 ZEOLITES

Zeolites are used in chemical industries for catalytic cracking. Zeolites are alumino silicates. Bentonites have also a number of industrial applications. They are phyllo alumino silicates. It has been revealed from the structure of zeolites that it has cages with pores of variable size. Pentasil family of high-silica zeolites (ZSM-5) has unique pore structures of intersecting channels. That is why ZSM-5 is being used for the removal of arsenic and heavy metals from an aqueous medium. The catalytic activity of zeolites is due to the association of hydroxyl groups with the aluminum ions. Synthetic and natural zeolites both are used as the catalyst. The mechanism of adsorption by zeolites is also explained as taking place by diffusion of reactant molecules through intracrystalline pores and chemisorption at the surface of zeolites. Surface area of the catalyst available for reaction and configuration of the diffusing species is the key factors during adsorption. Experimentally intraparticle diffusion has been established in removal of fluoride and arsenic by zeolites, clay, and clay minerals. Thus, diffusion as well as adsorption in the pores of the zeolites account for removal of toxic elements.

Zeolite are shape-selective catalysts. The reactants diffuse inwards to the active center located at pores. Geometric factors determine the shape selectivity of the catalysts. Transition metal oxides act as redox catalyst also. Some of the nontransition metal oxides such as Al_2O_3 and SiO_2 are also good catalysts. A large number of alumino silicate zeolitic acids containing Al_2O_3 and SiO_2 in different ratios are used as acid–base catalysts. Alumino silicate zeolitic acid can be synthesized with different ranges of Si/Al ratios. Thus, the activity of synthetic zeolite can be increased by varying the ratio of Si and Al.

Clay minerals may be classified according to the ratio of tetrahedral to octahedral sheets in one unit cell. Group is defined as according to charge density on the layer surface. Subgroup is defined according to the nature of octahedral sheet.

Clay minerals, in general, have a layered structure based on sheet with six-membered rings of silica tetrahedra, for example, pyrophyllite, beidellite, nontronite, saponite, iron saponite, and vermiculite. The montmorillonite mineral in particular has a layered structure of sheets of silica tetrahedra with Al or Mg octahedral, for example, (2:1). One octahedral sheet is sandwiched between two tetrahedral sheets. The clay minerals have a layered structure with six-membered rings of silica tetrahedral. This results due to the condensation of sheets of aluminum or magnesium octahedral. The clay minerals can be classified on the structural basis like 2:1 type, 1:1 type.

11.6.1 2:1 TYPE

Pyrophyllitic-talc group

Disubgroup	Pyrophyllite
Trisubgroup	Talc

Many groups and subgroups are present

- Smectite group or montmorillonite-saponite group (montmorrilonite minerals).
- Disubgroups like montmorrilonite, beidellite, nontronite.
- Trisubgroups like saponite, iron saponite.
- Vermiculite Group (vermiculite minerals).
- Disubgroup: vermiculite having aluminum in the interlayer.

- Trisubgroup: vermiculite having magnesium in the interlayer.
- Mica group: hydromuscovite, celadonite.

11.6.2 1:1 TYPE

1. Kaolinite serpentine group.
2. Kaolinite halloysite.
3. Non crystalline clay mineral: allophane, imogolite.
4. Interstratified clay: saponite, chalorite, tosudite.

11.6.2.1 REGULAR TYPE

11.6.2.1.1 Corrensite

This mineral is expandable and is swelling chlorite. Vermiculite–montmorillonite regular interstratifications of mica and an expandable mineral.

11.6.2.1.2 Pyrophyllite and Talc

The pyrophyllite–talc group has two subgroups.

a) Dioctahedral pyrophyllite.
b) Trioctahedral talc.

Pyrophyllite may have two crystal systems.

a) Monoclinic.
b) Triclinic.

Talc occurs in dolomite and serpentine rocks.

11.7 BENTONITE-MONTMORILLONITE

Montmorillonite is hydrated soft phyllosilicate, a member of smectite family having two tetrahedral sheets sandwiching a central octahedral sheet, distributed widely in the Rajmahal hills of Jharkhand and Hazaribagh district too. Because of the most important properties of these clays and its derivatives, they are considered as the clay minerals. Montmorillonite as a clay mineral has been discovered in Montmorillon, France, which contains high alumina with an expanding lattice. Rajmahal hills have a natural reserve of 568.36 million tons, a share of 0.17% with respect to the country as per the data of the Department of Mines and Geology, Government of Jharkhand. The mineral has a defined and definite structure. This mineral can be implied in academic studies and economic importance. Bentonites possess a three-layer structure. It has color absorption and remarkable cation exchange property.

Bentonites originated from weathering of volcanic ash in presence of water. Bentonites are mainly found close to surface to a depth of about 25 m. Many eminent geologists presented the evidence that most of the bentonite beds are being formed from volcanic minerals and found to be transported at a considerable distance in the atmosphere. They came to this conclusion based on the different factors like the purity of smectite, nonclay mineral present in it, and on the presence of shard textures in it. Bentonite deposits in alkaline lakes are formed very soon after the deposition, whereas other types are formed comparatively at a slower process. It is a highly colloidal plastic clay particularly found near Fort-Benton in the cretaceous beds of Wyoming in the USA. The origin of the bentonite is related to contemporaneous acid volcanic activity. In India, bentonites are found in Bihar, Jharkhand, Jammu and Kashmir, Tamilnadu, and Rajasthan. Bentonites of Rajasthan are a matter of interest for the researchers

11.7.1 ORIGIN AND OCCURRENCE OF BENTONITES

Wilbur C. Knight was the first to propose the term bentonite in 1898 as it is found in Fort Benton (Wyoming, USA). In the early part of 20th century, bentonite is mainly found in beds in cretaceous and tertiary rocks that are mainly originated from transported volcanic materials. Hannett told that the clay was formed by the alternation of volcanic ash. In 1926, Ross and Shannon observed that the clay was mainly composed of smectite clay mineral having swelling property. According to them, bentonite is a rock composed of crystalline clay-like structure formed by devitrification and accompanying chemical alternation of glassy igneous materials, a tuff or volcanic ash. Volcanic rocks form different minerals like biotite, kaolinite, quartz, pyroxene, and zircon. These clay minerals were found to have high birefringence, facile cleavage, texture of volcanic ash. The alternation of volcanic rock to montmorillonite requires humid climate conditions with stagnant water.

Bentonites are found to have the industrial features. Bentonite term can be applied to any plastic, colloidal, and swelling clay composed of smectite minerals of montmorillonite groups

Bentonite is a rock formed of highly colloidal and plastic clays composed mainly of montmorillonites, a clay mineral of the smectite group. It is a mixture of minerals. This material contains clay and nonclay minerals. Ross and Shannon concluded that essential clay minerals of bentonites such as montmorillonite and beidellite found to have the formula of [(Mg, Ca) O. $Al_2O_3.5SiO_2.nH_2O$][1] here n is 5–8 and $Al_2O_3.3SiO_2.nH_2O$ (n is 4). Sodium and calcium are the two main types of bentonites. Sodium bentonites are highly swelling, whereas calcium bentonites are nonswelling. The primitive factor for the formation of montmorillonites is the vast deposition of volcanic minerals. A high concentration of magnesium and silicate is largely responsible for the formation of montmorillonite. Bentonites are found formed by hydrothermal activity that occurs in Nevada, California, Spain, and other places. Bentonites are generally processed by breaking large pieces into smaller fragments followed by drying at low to moderate temperature to remove water present in it. The normal size of particles is generally 200 meshes that is equivalent to 70 μm. For modification of the type of exchangeable ions[2,3] in the crystal lattice structure process of beneficiation and removal of impurities like sand and others. Calcium bentonites are produced for the purposes like paint, special greases, etc. This process was applied for producing sodium montmorillonite by replacing calcium from the montmorillonite which is best applied for drilling mud. Organic clay bentonites are useful to produce paint, special grease. For this preparation, the inorganic exchangeable cations in the montmorillonite are exchanged with alkyl ammonium organic cation. The special products formed are commercially named as Bentone. In India, bentonites are found in Bhavnagar, Kutch of Gujrat and Balmer districts of Rajasthan are rich in bentonites. These bentonites of Rajasthan and Gujrat differ from the Rajmahal and Hazaribagh bentonite with respect to the source rocks, climate of source area, the environment of the basin of sedimentation. Research workers concluded that Rajmahal Bentonites are of high quality and be exploited for catalytic and adsorption properties. Different chemical and physical properties have been studied by scientists and research scholars. Through these studies, the following observations are found:

1. A small amount of Kaolinite and traces quartz are present in bentonites.
2. Dioctahedral montmorillonites kaolinite, limonite, and quartz are present in small amount in bentonites (Hathi Singh Ki Dhani Bentonites), the methods applied are DTA, electron micrography, and X-ray analysis.[4–9]
3. Researchers carried out extensive X-ray studies of bentonite from Balmer district of Rajasthan and observed that it contains Montmorillonite as a clay mineral and quartz as nonclay substance.
4. Cation exchange property of bentonites and softening of hard water have been studied successfully in the Chemistry Department of G.B. College, Naugachia under T.M. Bhagalpur University by Dr. Ashok Kumar Jha in research projects. Removal of some of the toxic metals by bentonites by adsorption and cation exchange has also been studied.
5. A theoretical molecular formula for the bentonite present in Santal Parganas of Jharkhand has been calculated.

11.7.2 THESE CLAY MINERALS HAVE THE FOLLOWING CHEMICAL COMPOSITION

The chemical formula of montmorillonite is $(Na, Ca)_{0.3}(Al, Mg)_2SiO_4O_{10}(OH)_2 . nH_2O$ and chemical composition of montmorillonite show various oxides, for example, SiO_2, Al_2O_3 Fe_2O_3, FeO, MnO, MgO, CaO, Na_2O, and K_2O and traces of TiO_2 . In chemical formula, tetrahedral sheet is occupied wholly by Si and no. of octahedral cations is 2. The percentage of Al_2O_3 varies from 4 to 19% approx. and percentage of SiO_2 is up to 60%, Fe_2O_3 varies from 7.8 to 9.5% and MnO varies from 2.6 to 3.3%. Cation exchange capacity of montmorillonite units in milli equivalent per 100 g ranges from 75.6 to 98.2. Generally, Rajmahal hill bentonites have C.E.C up to 89 m eq per 100 g.

Montmorillonite is generally basic in nature having pH 8.7–9.4.

Montmorillonite deposits are spread in Rajmahal hills in different colors and grades. Some deposits are white, green, and brown. Montmorillonite deposits having a brown color show the presence of ferric hydroxide in mineral.

The surfaces of montmorillonite are populated by oxygen atoms due to which reaction with other compounds is promoted. So bentonites dispersed in water become reactive. Bentonite solution in water behaves like colloidal solution. If NaCl solution is added, flocculation takes place and accordingly no swelling. This phenomenon is due to the repulsion of like ions associated with the clay and like ions of the suspending medium. Bentonites and their derivatives also find application as adsorbents of heavy metals, toxic elements, and fluoride. Bentonites also find applications an adsorbent of organic dyes. Effluents of dye after coloring fibers pass into water bodies thus adding toxic elements to water which is harmful for human beings and aquatic systems. Initial studies have shown those bentonites and their derivatives are useful in removal of organic dyes from an aqueous medium. The surface properties have been studied for the removal of many toxic properties from an aqueous medium.

Bentonites from all over the world have been utilized to remove heavy metals and fluoride from an aqueous medium by adsorption. Several adsorption isotherms have been proposed among which Freundlich and Langmuir isotherms are very common. Some adsorption isotherms are given below.

11.7.3 SURFACE PROPERTIES

Model	Assumption
Dubinin–Radush Kevich	Heterogeneous surface
Elovich	Kinetic-based model multilayer adsorption
Freundlich	Multilayer layer heterogeneous physisorption
Langmuir	Single layer homogeneous chemisorptions
Temkin	Intraparticle interactions chemisorptions Linear change in heat of adsorption
Brunauer–Emmett–Teller(BET)	Mono and multilayer surface adsorption area homogeneous
Frenkel–Halsey–Hill	Multiple layers assumed interlayer spacing and interaction
Toth	High to low dilutions, Heterogeneous, Single and multiple layer
Anderson	Adsorption area is lower in progressing layers. Finite adsorption layers. All concentration of pollutant

11.8 LANGMUIR'S THEORY OF ADSORPTION

In 1916, Langmuir postulated a theory of adsorption of gases on a solid surface. Langmuir suggested that adsorption consists of two processes.

i) Condensation of the molecules of the gas on the surface of the solid.
ii) Evaporation or desorption of the molecule from the surface of the adsorbent.

The solid surface becomes gradually covered, the rate of condensation becomes slow. The rate of desorption of the condensed molecules is slow at the beginning of the condensed molecules, but increase with the increasing number of occupied sites. At equilibrium two rates become equal.[10–16]

On the basis of the above postulates Langmuir derived the isotherm as:

$$\text{A(g)} + \text{S (surface)} \xrightarrow[k_2]{k_1} \text{AS}$$

where k_1 and k_2 are the adsorption and desorption constants, respectively.

Let θ be the fraction of surface sites occupied by the gas molecules. So $(1-\theta)$ is the unoccupied surface by the gas molecules.

μ is the number of molecules of gas striking 1 cm^2 of the surface per second. α is the proportion that adheres.

$\alpha\mu$ molecules adhere to each sq. cm of the surface per second.

$$\text{Rate of condensation} = (1-\theta)\,\alpha\mu$$

$$\text{Rate of desorption } \alpha \text{ area of the covered surface}$$

$$\text{Rate of desorption } \alpha\ \theta$$

$$\text{Rate of desorption} = V\theta, \text{ where } V \text{ is a constant}$$

At equilibrium,

$$\text{Rate of condensation} = \text{Rate of desorption.}$$

$$(1-\theta)\,\alpha\mu = V\theta \tag{11.1}$$

$$\text{or, } \theta\,(V + \alpha\mu) = \alpha\mu$$

$$\therefore \theta = \frac{\alpha\mu}{V + \alpha\mu} \tag{11.2}$$

$$\text{or, } \theta = \frac{1}{1 + \dfrac{V}{\alpha\mu}} \tag{11.3}$$

μ is proportional to the pressure of the gas.

$\mu = kP$, where k is proportionality constant.

$$\text{Hence, } \theta = \frac{1}{1+\dfrac{V}{\alpha kP}} \tag{11.4}$$

Let k_1 be equal to $\dfrac{\alpha k}{V}$

$$\text{or, } \theta = \frac{k_1 P}{1+k_1 P} \tag{11.5}$$

θ is proportional to the amount of gas x adsorbed by mass m of the adsorbent.

$$\frac{x}{m} \alpha \theta$$

$$\text{or, } \frac{x}{m} = k_2 \theta \tag{11.6}$$

$$\frac{x}{m} = \frac{k_2 k_2 P}{1+k_1 P} \tag{11.7}$$

This equation can be rewritten as:

$$\frac{P}{x/m} = \frac{1}{k_1 k_2} + \frac{P}{k_2}$$

This is the Langmuir adsorption isotherm equation.

A plot of $\dfrac{P}{x/m}$ against P gives straight line.

i) When the pressure is very low, only a small fraction of the surface is covered with molecules. θ is very small and so $(1 - \theta)$ is almost equal to unity.

$$\frac{x}{m} = k_1 k_2 P = kP$$

Thus at low pressure, the adsorbed amount is proportional to the pressure.

ii) When the pressure is high, θ is almost equal to unity

$$\text{From } \frac{x}{m} = \frac{k_1 k_2 P}{1+k_1 P}$$

$$k_1 P >> 1$$

$$\frac{x}{m} = \frac{k_1 k_2 P}{k_1 P} = k_2$$

Thus, at high pressure $\frac{x}{m}$ attains the limiting value, that is, the solid surface is covered completely with a single layer of gas molecules.

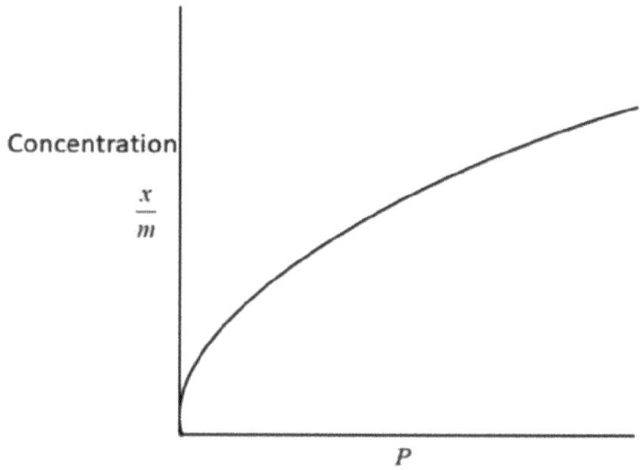

FIGURE 11.1 Plot of Langmuir adsorption isotherm.

The Langmuir adsorption isotherm thus explains the adsorbed amount of gas on the adsorbent surface. The amount of adsorbed gas per unit mass of adsorbent varies with pressure until the surface is covered with a single layer of molecules thus equilibrium exists between gas adsorbed on a surface and desorption and on this basis the isotherm equation has been derived. Langmuir isotherm explains the kinetics of heterogeneously catalyzed reactions.

Langmuir adsorption isotherm explains monolayer adsorption. The monolayer is formed when the strength of the bond between adsorbate–adsorbent molecules is stronger. The second next adsorbed molecule is just adjacent to the first and as a result of strong adsorbent–adsorbate interaction, only a monolayer is formed.

11.9 DIFFERENCE BETWEEN PHYSISORPTION AND CHEMISORPTION

i) Physical adsorption occurs at low temperatures and chemisorption occurs at all temperatures.
ii) Physisorption decreases with a rise in temperature whereas chemisorption increases with rise in temperature.

iii) The heat evolved in physisorption is low, whereas the heat evolved in chemisorption is high.
iv) Physisorption is reversible, that is, the adsorbed gas is released from the adsorbent surface by lowering the pressure at the same temperature, whereas chemisorption is reversible, that is, once the gas is adsorbed on the surface, the adsorbed gas cannot be recovered by lowering the temperature.
v) Physisorption is multilayer whereas chemisorption is monolayer.
vi) The adsorbate molecules are held by weak Van der Waals forces. So adsorbate molecules can be desorped easily or the activation energy of desorption is very low. The adsorbate molecules are held by valence bond so they cannot be desorped.

Freundlich isotherm refers to multilayer adsorption and Langmuir adsorption isotherm refers to monolayer adsorption. The adsorption of toxic elements from an aqueous medium follows Langmuir adsorption isotherm in some cases and in some cases Freundlich isotherm is followed.

11.10 BRAUNER, EMMETT, AND TELLER (BET ISOTHERM)

The Langmuir adsorption isotherm explains the formation of monomolecular layer. But multilayer adsorption also takes place. The theory of multilayer adsorption has been proposed by Brauner, Emmett, and Teller in 1938 popularly known as the BET theory. In the BET theory, it has been assumed that the solid surface possesses uniform sites. Adsorption at one site does not affect adsorption at other neighboring sites. Following are the postulates of BET theory.

i) This theory is an extension of Langmuir's theory of monolayer adsorption.
ii) The condensation forces are the principal forces in adsorption.
iii) The rate of evaporation from the first layer is equal to the rate of condensation on uncovered surface.
iv) The rate of evaporation from each succeeding layer is equal to the rate of condensation on the preceding layer.
v) The heat of adsorption is involved in each of the equilibrium evaporation rate constant.
vi) The energy of adsorption in the first layer is assumed to be constant and the energy of adsorption in succeeding layers is assumed to be the same, the energy of liquefaction of gases.

11.10.1 DERIVATION

Let S_0, S_1, S_2,…,S_i be the surface area of the adsorbent covered by 0, 1, 2,…, i layers of the adsorbed molecules.

At equilibrium, S_0 remains constant. The rate of condensation on the empty surface is equal to the rate of evaporation from the first layer.

The rate of adsorption is proportional to the pressure of the gas αPS_O.

Rate of condensation = a_1PS_O, where a_1 is a constant.

Rate desorption is proportional to the covered surface area.

From Boltzmann's exponential function

$$k_2 = b_1 e^{-E_1/RT}$$

where b_1 is a constant and E_1 is the heat of adsorption of the first layer.

Rate of desorption $= b_1 S_1 e^{-E_1/RT}$

$$\text{At equilibrium, } a_1 PS_O = b_1 S_1 e^{-E_1/RT} \quad (11.8)$$

S_1 can change by condensation on the empty surface, by desorption from the first layer, condensation on the first layer and desorption from the second layer

$$a_2 PS_1 + b_1 S_1 e^{-E_1/RT} = b_2 S_2 e^{-E_2/RT} + a_1 PS_O \quad (11.9)$$

From Equation 11.8, we have

$$a_2 PS_1 = b_2 S_2 e^{-E_2/RT} \quad (11.10)$$

where a_2, b_2, and E_2 have their usual meanings.

Thus, $a_3 PS_2 = b_3 S_3 e^{-E_3/RT}$

$$a_i PS_{i-1} = b_i S_i e^{-E_i/RT} \quad (11.11)$$

Whole surface area of the catalyst (A) is given by

$$A = \sum_{i=0}^{\infty} S_i$$

The two-volume adsorbed is given by:

$$V = V_O \sum_{i=0}^{\infty} i.S_i,$$

where V_O = volume of the gas adsorbed on 1 cm^2 of the adsorbent.

$$\frac{V}{V_O.A} = \frac{\sum_{i=0}^{\infty} i.S_i}{\sum_{i=0}^{\infty} S_i} = \frac{V}{V_m} \tag{11.12}$$

Let us assume, $E_2 = E_3 = E_i = E_l$

where E_l is heat of liquefaction

$\frac{b_2}{a_2} = \frac{b_3}{a_3} = \frac{b_4}{a_4} = \frac{b_i}{a_i} = g$, where g is a constant

$S_1, S_2, S_3 \ldots\ldots S_i$ can be expressed in terms of S_O as

$$S_1 = yS_O \tag{11.13}$$

where $y = \frac{a_1}{b_1} P.e^{E_1/RT}$

$$S_2 = xS_1 \tag{11.14}$$

where $x = \frac{P}{g} e^{E_1/RT}$

$$\therefore\ S_2 = xyS_O \tag{11.15}$$

$$S_3 = xS_2 = xxS_1 = x^2 S_1$$

$$\text{or, } S_i = xS_{i-1} \tag{11.16}$$

$$= x^2.\frac{y}{x} S_O = c.x^{iS_O}$$

where $c = \frac{y}{x} = \frac{a_1 \cdot g}{b_1}.e^{(E_1 - E_l)/RT}$

Equation 11.12 can be written as

$$\frac{V}{V_m} = \frac{\sum_{i=0}^{\infty} i.S_i}{\sum_{i=0}^{\infty} S_i} = \frac{\sum_{i=0}^{\infty} i.S_i}{S_O + \sum_{i=1}^{\infty} S_i} = \frac{\sum_{i=o}^{\infty} i.c.x^i.S_O}{S_O + \sum_{i=1}^{\infty} cx^i.S_O} \tag{11.17}$$

$$\text{or, } \frac{V}{V_m} = \frac{cS_O \sum_{i=o}^{\infty} i.x^i}{S_O(1 + c\sum_{i=1}^{\infty} x^i)} \tag{11.18}$$

Now, $\sum_{i=1}^{\infty} x^i = \frac{x}{1-x}$ and $\sum_{i=0}^{\infty} i.x^i = \frac{x}{(1-x)^2}$

$$\frac{V}{V_m} = \frac{c.\frac{x}{(1-x)^2}}{[1+c.\frac{x}{(1-x)}]} = \frac{c.x}{(1-x)(1-x+cx)}$$

$$\text{or, } \frac{V}{V_m} = \frac{c.x}{(1-x)[1+(c-1)x]} \tag{11.19}$$

Rearranging Equation 11.19, we obtained the following equation:

$$\frac{x}{(1-x)V} = \frac{1}{cV_m} + \frac{(c-1)x}{cV_m} \tag{11.20}$$

where V_m = volume of the gas absorbed when the entire adsorbent surface is covered with a unimolecular layers. V is the total volume of the gas adsorbed at pressure P, is a constant depending on the nature of the gas.

When $\frac{x}{(1-x)V}$ is plotted against x, a straight line is obtained. The slope of straight line is equal to $\frac{(c-1)}{cV_m}$ and the intercept is equal to $\frac{1}{cV_m}$.

From this graph V_m is thus known.

The number of moles of the gas is known, surface area of the adsorbent may be calculated.

BET equation may be represented as:

$$\frac{P}{V(P_O - P)} = \frac{1}{V_m.c} + \frac{c-1}{V_m.c}.\frac{P}{P_O}$$

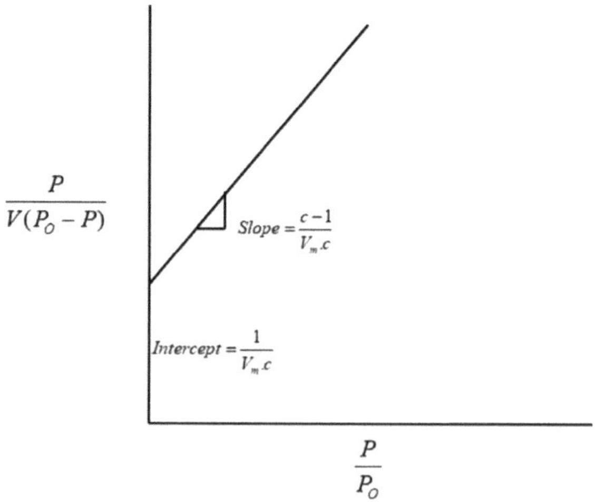

FIGURE 11.2 BET plot for the adsorption of N_2 on silica gel at −180°C.

The experimental data have been checked in the case of the adsorption of gases on solid adsorbents. In the case of adsorption of N_2 on silica gel, a good straight line is obtained.

At very high pressures deviation occurs.

The BET equation covers monolayer and multilayer isotherms. If $E_1 >> E_l$, the BET equation reduces to Langmuir adsorption isotherm.

Brauner observed that on plotting $\frac{P}{P_O}$ versus amount adsorbed gives different types of adsorption isotherm.

P_O represents saturation pressure and P is the measured pressure. $\frac{P}{P_O}$ represents the relative pressure.

The shape of the graphs depends on the nature, pore space, and surface area available.

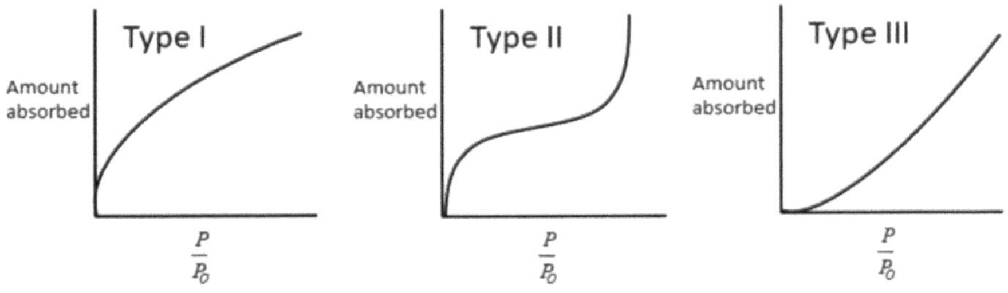

An isotherm for chemical adsorption is expressed by type I (Langmuir type). But adsorption by Van der Waals, forces account for multilayer adsorption types II and III.

The adsorption of the gases such as nitrogen and argon correspond to the behavior shown in type II isotherm.

11.10.2 SURFACE AREA OF ADSORBENT

The surface area includes the geometrical area, area of the capillaries, pores, kinks, and steps on the surface of the solid. The specific surface area is defined as the surface area per unit mass.

The product of the number of molecules in the monolayer and the molecular cross-section gives the surface area. Emett and Teller proposed the formula of surface area as

Let ρ be the density and M be the molar mass of the adsorbate. V is the volume occupied by a single molecule.

$$\rho = \frac{M}{V} = \frac{M}{N.V}$$

$$\text{or, } V = \frac{M}{N.\rho} \tag{11.21}$$

Molecule may be treated as spherical. Volume of the sphere is given as: $V = \frac{4}{3}\pi r^3$, from Equation 11.21, $\frac{4}{3}\pi r^3 = \frac{M}{N.\rho}$

$$\text{or, } r = \left[\frac{3M}{4\pi N\rho}\right]^{\frac{1}{3}} \tag{11.22}$$

If a is the area of cross section of a molecule

$$a = \pi r^2 = \pi\left[\frac{3M}{4\pi N\rho}\right]^{\frac{2}{3}} \tag{11.23}$$

This formula does not take into account the nature of packing at the solid surface. Void volume in the lattice has also not been considered.

Method: The supplied sample is made free of water or any other contaminant. It is heated in a vacuum. The temperature is selected such that the lattice of the sample is not disturbed. A small, for example, 0.5 g of sample is enough to determine the BET surface area. Now the sample is taken in a glass cell and analyzed by the instrument. This technique is used for the surface area of nanoparticles and metal-organic frameworks. The BET instrument applied is analytical (micromeritics Gemini 2375 and Gemini V) in which nitrogen is used as an adsorbate.

"Zetasizer lab" particle size and zeta potential data analyzer are used for measuring particle size from 0.3 nm to10 μm. Malvern panalytical is the supplier.

11.11 ADSORPTION FROM SOLUTION

Adsorption of a solute from solution on the solid surface is explained by Gibbs adsorption isotherm. The concentration of a solute at the surface of a solution differs from that in the bulk. When the surface tension of the solute is lower than liquid, it has the tendency to get accumulated at the surface, and thus surface tension decreases.

Let us consider a system of two components. The Gibbs free energy is written as

$$G = n_1\mu_1 + n_2\mu_2 \tag{11.24}$$

where n_1 and n_2 are the number of moles and $\mu_1\,\mu_2$ are the chemical potentials of the two components, respectively

$$G = n_1\mu_1 + n_2\mu_2 + \gamma\sigma \tag{11.25}$$

where Y is the surface energy and σ is the surface area. Differentiating Equation 11.25, we have:

$$dG = n_1 d\mu_1 + \mu_1 dn_1 + n_2 d\mu_2 + \mu_2 d\mu_2 + \gamma d\sigma + \sigma d\gamma \tag{11.26}$$

Free energy of the system depends on five independent variables T, P, n_1, n_2 and σ

$$G = f(T, P, n_1, n_2\sigma) \tag{11.27}$$

Differentiating Equation 11.27, we have

$$dG = \left(\frac{\delta G}{\delta T}\right)_{P,n_1,n_2,\sigma} .dT + \left(\frac{\delta G}{\delta P}\right)_{T,n_1,n_2,\sigma} .dP + \left(\frac{\delta G}{\delta n_1}\right)_{T,P,n_2,\sigma}$$

$$.dn_1 + \left(\frac{\delta G}{\delta n_2}\right)_{T,P,n_1,\sigma} .dn_2 + \left(\frac{\delta G}{\delta \sigma}\right)_{T,P,n_1,n_2} .d\sigma \tag{11.28}$$

$$dG = -SdT + VdP + \mu_1 dn_1 + \mu_2 dn_2 + \gamma d\sigma \tag{11.29}$$

where $\left(\frac{\delta G}{\delta n_1}\right)_{T,P,n_2,\sigma} = -\mu_1, \left(\frac{\delta G}{\delta \sigma}\right)_{T,P,n_1,n_2} = \gamma$

$$\left(\frac{\delta G}{\delta n_1}\right)_{T,P,n_2,\sigma} = -\mu_1, \left(\frac{\delta G}{\delta \sigma}\right)_{T,P,n_1,n_2} = \gamma$$

At constant temperature and pressure:

$$(dG)_{T,P} = \mu_1 dn_1 + \mu_2 dn_2 + \gamma d\sigma \tag{11.30}$$

Form Equations 11.26 and 11.30:

$$n_1 d\mu_1 + n_2 d\mu_2 + \sigma d\gamma = 0 \tag{11.31}$$

Expression for the bulk of the liquid is:

$$n_1^o d\mu_1 + n_2^o d\mu_2 = 0 \tag{11.32}$$

where n_1^o and n_2^o are the number of moles of the liquid and the solute in the bulk phase, respectively.

At equilibrium, the chemical potential of each component in the bulk and the surface must be the same.

From Equation 11.32:

$$n_1^o d\mu_1 = -n_2^o d\mu_2$$

$$d\mu_1 = -\frac{n_2^o}{n_1^o} d\mu_2$$

Putting the value of $d\mu_1$ into Equation 11.31, we have:

$$-n_1 . \frac{n_2^o}{n_1^o} d\mu_2 + n_2 d\mu_2 + \sigma d\gamma = 0$$

$$\text{or, } \left(n_2 - \frac{n_1 . n_2^o}{n_1^o} \right) d\mu_2 + \sigma d\gamma = 0$$

$$\text{or, } \frac{-d\gamma}{d\mu_2} = \frac{n_2 - \left(\frac{n_1 . n_2^o}{n_1^o} \right)}{\sigma} = \Gamma_2 \tag{11.33}$$

where n_2^o is the number of mole of solute 2 associated with n_1^o, the number of moles of liquid 1 in the bulk phase. Thus, $n_2 - \left(\frac{n_1 . n_2^o}{n_1^o} \right)$ is the excess amount of the solute present in the liquid surface. Thus, $\frac{n_2 - \left(\frac{n_1 . n_2^o}{n_1^o} \right)}{\sigma}$ is the excess concentration of the solute per unit surface area which is expressed by Γ_2. Thus, $\Gamma_2 = -\frac{d\gamma}{d\mu_2}$.

From thermodynamics, $\mu_2 = \mu_2^o + RT \ln a_2$

where μ_2 is the chemical potential of solute 2

Differentiating the above equation, $d\mu_2 = RTd \ln a_2$

$$\Gamma_2 = -\frac{1}{RT} . \frac{d\gamma}{d \ln a_2} = -\frac{a_2}{RT} . \frac{d\gamma}{da_2} \tag{11.34}$$

For dilute solution, activity is replaced by concentration:

$a = fC$, where a is activity, f is activity coefficient, and C is concentration
$a = C$ as $f = 1$

$$\Gamma_2 = -\frac{C_2}{RT}\cdot\frac{d\gamma}{dC_2} \tag{11.35}$$

This is known as Gibbs adsorption isotherm.

When the adsorption of a surfactant takes place at the interface, the concentration at the surface is higher than that in the bulk phase. Equation 11.35 shows the relation between surface tension and surface excess concentration.

If the surface excess concentration Γ_2 is positive, solute lowers the surface tension. For a solute raising the surface tension, Γ_2 is negative. Inorganic salts have lower concentration in the surface than in the bulk aqueous solution, whereas soaps have higher concentration in the surface than in bulk aqueous phase.

It is experimentally known that solute reduces the surface tension of water. This is why charcoal and silica gel adsorbs most of the solutes from aqueous solutions. A solute is readily adsorbed from an aqueous solution by a suitable adsorbent.

11.12 SURFACE STRUCTURE

Atoms on the solid surface experience different forces than in the bulk. Atoms on the surface are not fixed in a rigid manner as a result of which some mobility is developed on the surface atom during adsorption of gas on the surface. The crystal grows from a melt, solution, or vapor phase, and so the growing crystals have terraces, steps, and kinks. During growth of the crystal, an atom or an ion strikes the surface and get attached to the surface. It faces interaction with different atoms simultaneously and thus deposits here resulting in crystal growth. The arrangement of the atoms on the surface can be studied with low energy electron diffraction (LEED) because X-ray diffraction cannot determine the structure precisely. X-ray diffraction is more accurate for bulk solid than surface solid.

11.12.1 LEED METHOD

High-energy electrons penetrate many layers of atoms which give rise to diffraction patterns. Low energy electrons are reflected from the surface atoms. The diffraction is produced by few layers of surface atoms. If diffraction is produced from scattering of the surface layer, a characteristic diffraction pattern is obtained. The diffraction pattern is produced by the constructive interference of electrons. This is similar to three-dimensional diffraction.

Penetration of the low-energy electrons change with voltage. It is possible to locate the position of the atoms on the surface of a crystal. Thus, the position of the host as well as that of impurity or defect on the surface can be found out.

In the LEED instrument, a narrow beam of monochromatic electrons from an electron gun is allowed to strike the surface of a single crystal at the center of curvature of a fluorescent screen. The beam incident normal to the surface will be scattered back to produce the diffraction spots. Nickel-plated tungsten of spherical shape grid is placed in front of the screen, as shown in Figure 11.3.

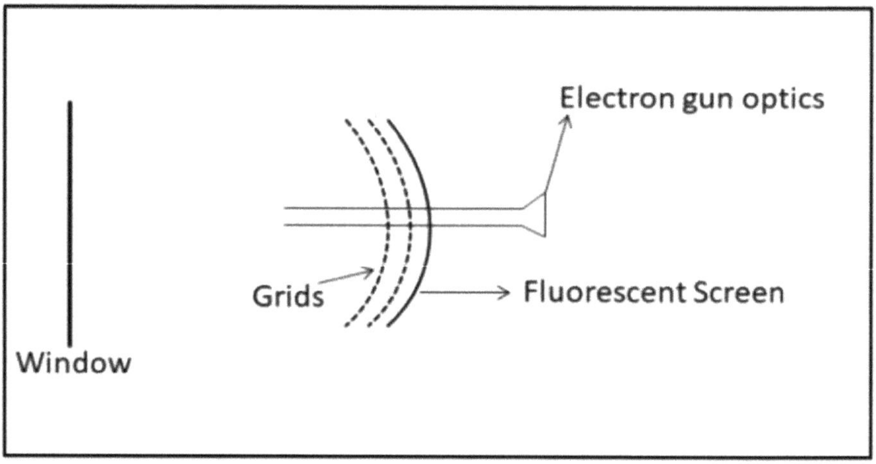

FIGURE 11.3 Sketch of LEED instrument.

The electron beam energy has been adjusted within the range 10–500 ev. Whole of the chamber is maintained at a vacuum of 10^{-10} torr (pressure) which is created with a view to make the surface free from any contamination and to avoid energy loss due to collision with gas molecules. The distance between the linear array of the surface atoms may be calculated from the geometry of the instrument.

After attaining a high vacuum, the surface of crystal inside the chamber is cleaned. The gas molecules cover the surface within a few hours. Within this time, a diffraction pattern of the clean surface is obtained. The diffused spots indicate the presence of impurity atoms present on the surface.

The presence of steps and kinks on the surface is thus identified from the LEED method.

Scanning tunneling microscopy and Auger electron spectroscopy along with LEED have been used to investigate doping on the crystal surface. In total, 0.5% (weight) niobium (Nb) doped $SrTiO_3$ is investigated by this method. Steps and kinks are known from this method.[25,26]

Thus, LEED provides all kinds of qualitative information, intensity of a spot, periodic structure, and a lot of imperfection, for example, steps, kinks, domains, and point defects. Random or regular arrangement may be identified by inspection of the LEED pattern. The defects due to static or thermal disorder, linear defects, steps between terraces, planer defects, and bulk defects are also known from the LEED pattern.[27]

Recently, tensor LEED has taken the place of LEED in 1885 which allows fast calculation of intensities for structures. The diffuse LEED technique known as DLEED has also been discovered. This allows the retrieval of the local structure in the case of disordered adsorption. So space images can be obtained directly through DLEED.[28]

One new instrument for spot profile analysis of electron diffraction (SPA-LEED) has made the analysis very easy. This instrument works with a transparent phosphor screen for measuring intensity. The scanning range covers 30° in this instrument which makes both qualitative and quantitative analysis.[29–31]

The simplest basic explanation of LEED is that a single crystal is bombarded with a collimated beam of low-energy electrons and diffraction electrons are observed on a fluorescent screen, as shown in Figure 11.3.

11.13 TEMKIN ISOTHERM

According to Temkin isotherm, heat of adsorption decreases with increase in coverage of the surface of the adsorbent. Uniform distribution of binding energies has been observed.

Temkin isotherm model is given by the following equation:

$$q_e = B_T \ln K_T + B_T \ln C_e$$

where $B_T = \frac{RT}{b}$, T is the absolute temperature, R is the universal gas constant, B_T is Temkin constant. Temkin constant is obtained by plotting q_e versus ln C_e. K_T(L/g^{-1}) is the equilibrium binding constant.

Temkin model has been applied in the adsorption of cadmium ion onto nonzero valent iron and in the adsorption of Pb(II) by chemically modified citrus limettioides peel.[32,33]

11.14 ELOVICH MODEL

This isotherm model is also applicable to multilayer adsorption with the assumption that adsorption sites increase exponentially with adsorption. The kinetics of chemisorption of gas on the solid surface have been explained with the help of this model also.

The linear form of the equation is

$$\ln\frac{q_e}{C_e} = \ln K_e q_m - \frac{q_e}{q_m}$$

Elovich constant is calculated by plotting $\ln\frac{q_e}{C_e}$ versus q_e. Intercept gives the value of $\ln K_e q_m$ and slope gives the value of $\frac{1}{q_m}$. From both slope and intercept, Elovich maximum adsorption capacity (q_m) and Elovich constant K_e are known.

11.15 LAGERGREN MODEL

Lagergren first-order model is expressed as:

$$\frac{dq_t}{dt} = k_1\left(q_e - q_t\right) \tag{11.36}$$

where q_e stands for adsorption capacity at equilibrium, q_t stands for adsorption capacity at time *t*. k_1 is the rate constant of pseudo-first-order adsorption. From Equation 11.36, we have

$$\frac{dq_t}{\left(q_e - q_t\right)} = k_1 dt,$$

Integrating, $\int \frac{dq_t}{q_e - q_t} = k_1 \int dt$

$$\text{or, } -\ln(q_e - q_t) = k_1 t + C \tag{11.37}$$

where C is the integration constant

$$\text{At } t = 0,\ C = -\ln q_e$$

$$\text{or, } -\ln(q_e - q_t) = k_1 t - \ln q_e$$

$$\text{or, } \ln(q_e - q_t) = \ln q_e - k_1 t$$

$$\text{or, } \log(q_e - q_t) = \log q_e - \frac{k_1}{2.303} t \tag{11.38}$$

A plot of $\log(q_e - q_t)$ versus t gives a straight line. The intercept is equal to $\log q_e$ and slope is equal to $\frac{k_1}{2.303}$. Thus k_1 and q_e can be calculated from the graph.

Lagergren second-order rate equation can be expressed as follows:

$$\frac{dq_t}{dt} = k_2 (q_e - q_t)^2 \tag{11.39}$$

where k_2 is the rate constant for a pseudo-second-order adsorption.

The integrated Equation 11.40 becomes equal to $\frac{t}{q_t} = \frac{t}{q_e} + \frac{1}{k_2 q_e^{\,2}}$

A plot of $\frac{t}{q_t}$ versus t gives a linear graph.

Both the plots of $\log(q_e - q_t)$ versus t and $\frac{t}{q_t}$ versus t show the applicability of either first order or second order.

The order of the reaction is defined as the sum of power of concentration terms used in the rate equation. But the order of a reaction may be altered by increasing or decreasing one of the reactants. If by changing concentration of reactants, the rate becomes independent of concentration, the order becomes zero. Such a reaction is known as pseudo-first-order reaction.

The pseudo-first-order and second-order reactions of Lagergren have been utilized for adsorption of fluoride, arsenic, and heavy metals onto the adsorbents, such as bentonites, agricultural residues, and zeolites.[34–36]

11.16 DUBININ–RADUSHKEVICH ISOTHERM

Dubinin–Radushkevich isotherm[37–38] is applicable to intermediate range of adsorbate concentrations. This model is semiempirical in nature. It also considers multilayer adsorption considering Van der Waal's forces. This model is temperature dependent and so adsorption data at different temperatures are plotted versus the square of Polanyi potential. Thus, Dubinin–Radushkevich estimates the characteristics porosity and free energy of adsorption. This isotherm has been expressed in the linear form as follows:

$$\ln q_e = \ln q_m - K \in^2$$

where K is a constant related to the adsorption energy, q_m is the monolayer saturation capacity (mg.g^{-1}), and $\in$ is the Polanyi potential

$$\in = RT \ln(1 + \frac{1}{C_e})$$

The slope of the plot of ln q_e versus $\in^2$ gives K (mol^{2}kJ^{-2}) and intercept gives the value of adsorption capacity q_m (mg.g^{-1})

E is the mean free energy of adsorption. When 1 mole of ion is transferred from infinity in solution to the surface of the solid, the free energy change is known as the mean free energy of adsorption E.

$$E = \sqrt{2K}$$

q_e (mg.g^{-1}) and C_e (mg.L^{-1}) are equilibrium adsorbate concentrations in the solid and liquid phases.

11.17 INTRAPARTICLE DIFFUSION

Intraparticle diffusion happens in a solid particle. Intraparticle diffusion parameter is calculated from adsorbent mass, particle size, and initial adsorbate concentration.

Intraparticle diffusion involves the transport of solute molecules from the aqueous phase to the solid surface by diffusion. This is the rate-determining step and hence a slow process. The empirical equation of this diffusion is shown as:

$$q_t = Kt^{0.5} + C$$

where C is the intercept and K is the intraparticle diffusion rate constant. The plot of q_t against $t^{1/2}$ show multi linear lines. This indicates that slop of the line gives the value of K and C is the intercept. The physical interpretation of C is that C shows thickness of the boundary layer. Intraparticle diffusion plays an important role in the uptake of heavy toxic metals, for example, chromium and lead by adsorbent.

If intraparticle diffusion is the only rate-determining step, the straight line passes through the origin on plotting q_t versus $t^{1/2}$. It also happens that surface adsorption and intraparticle diffusion take place simultaneously.

11.18 HARKIN–JURA ISOTHERM

This isotherm is also applicable to multilayer adsorption which is expressed by the equation given below:

$$\frac{1}{q_e^2} = \frac{B}{A} - \left(\frac{1}{A}\right)\log C_e$$

where B and A are Harkin–Jura constants. When $\frac{1}{q_e^2}$ is plotted against log C_e, Harkin–Jura constants B and A are calculated. The slope and intercept give the value of $\frac{1}{A}$ and $\frac{B}{A}$, respectively. Some researchers have found that Harkin–Jura isotherm shows a good fit to the experimental data of adsorption of organic dyes on Bentonite and derivative of Bentonites.

11.19 JOVANOVIC ISOTHERM

The Jovanovic isotherm is based on the assumption that monolayer adsorption takes place like that of Langmuir model. In addition, some mechanical contact may take place between adsorbate and adsorbent.

The isotherm is given as:

$$\ln q_e = \ln q_m - kC_e$$

where q_e is the amount of adsorbate in the adsorbent at equilibrium (mg. g^{-1}), q_m is maximum amount of adsorbate. When $\ln q_e$ is plotted versus C_e, slope gives the value of k, Jovanovic constant. This isotherm also explains some of the adsorption on solid surfaces.

KEYWORDS

- **Lagergren**
- **bentonite**
- **Temkin**
- **isotherm**
- **solid surface**

REFERENCES

1. Gupta, A.; Kumar, M.; Jha, A. K.; Mishra, B. K.; Choudhary, J.; Mishra, B. Chemical Composition and Molecular Formula of Rajmahal Bentonite (Jharkhand). *Asian J. Chem.* **2011,** *23* (12), 5491–5494.
2. Jha, A K.; Jha, A. K.; Mishra, A. K.; Kumari, V.; Mishra, B. Softening of Hard Water by Bentonite Mineral. *Asian J. Water Environ. Pollut.* **2011,** *8* (4), 93–96.
3. Jha, D.; Mishra, B. K.; Mishra, S. S.; Jha, A. K. Bentonite clay Minerals of Rajmahal Hills. *J. Haematol. Ecotoxicol.* **2010,** *5* (1), 1–7.
4. Jha, A. K.; Mishra, B. Removal of Fluoride by Bentonite Minerals of Rajmahal Hills *J. Indian Chem. Soc.* **2012,** *89* (4), 519–521.
5. Majumdar, S.; Jha, A. K.; Pranita; Mishra, K. K. Characterization of Montmorillonite Minerals. *JETIR* 2019, *6*, 3.
6. Jha, A. K. Bentonite Minerals And It's TGA, DSC And PXRD Studies. *J. Indian Chem. Soc.* **2016,** *93*, 437–442.
7. Jha, A. K. Bentonites for Chemical Industries. *J. Indian Chem. Soc.* **2018,** *95* (1), 35–40.
8. Rao, C. N. R.; Gopalakrishnan, J. *New Directions in Solid-State Chemistry*, 2nd ed.; Cambridge University Press, 1997.
9. Alberti, G.; Constantino, U. *Intercalation Chemistry*; Whittingham, M. S., Jacobson, A.J., Eds.; Academic Press: New York, 1982.
10. Jha, A. K. Adsorption of Cr (VI) and Arsenic onto Bentonite. *Chem. Sci. Rev. Lett.* **2014,** *3* (12), 1182–1189.
11. Removal of Fluoride from Aqueous Medium by Using Low Cost Bentonite of Rajmahal and Hazaribagh, Jharkhand. *Indian J. Environ. Protect.* Oct. **2020,** *40* (10).
12. Kinetic and Adsorption Study for Removal of Arsenic from Aqueous Medium by Low Cost Bentonite of Rajmahal Hills and Hazaribagh, Jharkhand. *Nat. Environ. Pollut. Tecnol.* **2020,** *19* (5), 1847–1852.
13. Powdered X-Ray Diffraction, FTIR, TGA and DTA Studies of Montmorillonite Derivatives. *J. Indian Chem.Soc.* Sept **2020,** *97*, 9b, 1604–1608.
14. Jha, A. K.; Majumder, S. Removal of Cr and Mn from Aqueous Medium Using Bentonites and Their Derivatives. *J. Chem. Sci.* **2020,** 132–135.

15. Jha, A. K.; Anjana. Studies on Biosorption of Pb(II) by Maize Stem and Rice Hus Powder. *Rasayan J. Chem.* **2021,** *14* (2), 937–942.
16. Jha, A. K.; Kiran, K.; Anjana, K.; Pranita. Removal of As(III) with (Cynodon Dactylon) Green Dub and Orange Peel from Aqueous Medium. *Rasayan J. Chem.* **2021,** *14* (2), 1072–1081.
17. Midya, A.; Mandal, P.; Rubi, K.; Chen, R.; Wang, J. S. Large Adiabatic Temperature and Magnetic Entropy Change in $EuTiO_3$. *Phys. Rev. B.* 2016.
18. Mady, T. E.; Yates, Jr, J. T.; Sandstrom, D. R.; Voorhoeve, R. J. H. *Treatise on Solid-State Chemistry*, Vol 6B; Plenum Press: New York, 1976.
19. Pauling L. *Proc. Roy. Soc. London A* **1949,** *196*, 343.
20. Desiraju, G. R. *Proc. Indian Acad. Sci Chem. Sci.* **1984,** *93*, 407.
21. Desiraju, G. R. *Crystal Engineering: The Design of Organic Solids*; Elsevier: Amsterdam, 1989.
22. Dowden, D. A. *Catal. Rev.* **1972,** *5*, 1.
23. Kulkarni, G. U.; Rao, C. N. R. *Catal. Lett.* **1991,** *9*, 427.
24. Centi, G.; Trifiro, F.; Ebner, J. R.; Fran Chetti, V. M. *Chem. Rev.* **1988,** *88*, 55.
25. Russel, B. C.; Castell, M. R. Reconstructions on the Polar $SrTiO_3$ (110) Surface: Analysis Using STM, LEED and AES.
26. Fujihaga, Y. A Surface Analysis of Cu-Au Alloy by LEED and Auger Electron Spectroscopy. *Surf. Sci.* **1997,** *64* (2), 751–770.
27. Henzler, M. LEED Studies of Surface Imperfections. *App. Surf. Sci.* **1982,** *11*, 450–469.
28. Heinz, K. LEED and DLEED as Modern Tools for Quantitative Surface Structure Determination. *Rep. Progress Phys.* **1995,** *58* (6), 637.
29. Scheithauer, U.; Meyer, G.; Henzler, M. A New LEED Instrument for Quantitative Spot Profile Analysis. *Surf. Sci.* **1986,** *178* (1–3), 441–451.
30. Over, H.; Kleinle, G.; Ertl, G.; Moritz, W.; Ernst Heinz, K-H. A LEED Structural Analysis of the CO(1010) Surface. *Surf. Sci.* **1991,** *254* (1–3), L 469–L474.
31. Feder, R.; Player, H.; Bauer, P.; Miller, N. Spin Polarization in Low Energy Electron Diffraction: Surface Analysis of Pt (III). *Surf. Sci.* **1981,** *109* (2), 419–434.
32. Sudha, R.; Srinivasan, K. Thermodynamic and Kinetic Studies on Pb (II) Removal from Aqueous Solution by Chemically Modified Citrus Limettioides Peel. *IJEP* **2015,** *35* (4), 300–311.
33. Ayawei, N.; Ebelegi, A. N.; Wankasi, D. *Modelling and Interpretation of Adsorption Isotherms*; Academic Editor: Whenshan Guo, Sept 5, 2017.
34. Kumari, A.; Jha, A. K.; Pranita; Kumari, K. Adsorptive Removal of Chromium Using Rice Husk and Maize Stem Powder. *Int. J. Latest Transitions Eng. Sci.* **2020,** *8* (4), 1–5.
35. Singh, B.; Jha, A. K. Defluoridation of Water by Montmorillonite Minerals. *Int. J. Latest Transition Eng. Sci.* **2020,** *8* (4), 6–8.
36. El Nemr, A. Potential of Pomegranate Husk Carbon for Cr (VI) Removal from Wastewater: Kinetic and Isotherm Studies. *J. Hazard. Mater.* **2009,** *161*, 132–141.
37. Enhanced Photocatalytic Activity of Solar Light Driven ZnS Composites Incorporated with Phosphotungstic Acid, Wenli Zhu, Solid-State Sciences, 6 September 2020, 1064.
38. Sujith, P. et al. Density Study of Electronic and Optical Properties of Quasi-One-Dimensional Metal Tin Sulfides Functional MSnS3 (M = Hf, Zr). *Solid-State Sci.* April **2021,** 106608.

Examination Problems

CHAPTER 2: QUESTIONS

i) What are the differences between TGA and DTA?
ii) Explain all the weight losses in the case of TGA of montmorillonite mineral.
iii) How does PXRD help characterize the clay samples?
iv) Explain OH, Si–O and Al–O peaks in F.T.I.R of clay minerals, for example, kaolinite and montmorillonite.
v) Explain the difference between SEM and TEM.
vi) How does SEM help analyze the adsorption?

CHAPTER 3: OBJECTIVE

1. In the closest packing ABCABC, z is the number of atoms in the unit cell. Find the number of tetrahedral voids in the unit cell.

(a) 1z
(b) 2z
(c) $z/2$
(d) $z/3$

2. When Y_2O_3, BaO_2, and CuO are used together to prepare a semiconductor $Y_2Ba_2Cu_3O_7$, the ratio molar ratio should be

(a) 1, 2, 4
(b) 1, 2, 3
(c) 3, 2, 1
(d) 1, 1.5, 2.5

3. A cubic solid lattice is crystallized from the elements A, B, and C. A atoms occupy the corners, B atoms the cube centers, and C atoms the edges. The formula of the compound is

(a) ABC
(b) ABC_2

(c) ABC_3
(d) AB_3C

4. The tetrahedral interstitial site is formed by

(a) Four spheres
(b) Partly same and different spheres
(c) Four spheres, the centers of which give rise to a regular tetrahedron
(d) None

5. If hexagonal closest packing is formed by anions A and 2/3rd octahedral voids are occupied by cations C, the general formula of the compound will be:

(a) CA
(b) CA_2
(c) C_2A_3
(d) C_3A_2

6. The coordination number ratio of cation and anion in Fluorite CaF_2 and CsCl are given as

(a) 8:4 and 6:3
(b) 6:3 and 4:4
(c) 8:4 and 8:8
(d) 4:2 and 2:4

7. For a zinc blende unit cell, the number of formula units of ZnS is:

(a) 1
(b) 2
(c) 4
(d) 8

8. The greater Schottky defect is shown by

(a) CaF_2
(b) ZnS
(c) AgCl
(d) CsCl

9. The diffraction of x-rays by a crystal is possible only when

(a) $\lambda \leq 2d$
(b) $\lambda \geq 2d$
(c) $\lambda > 2d$
(d) $\lambda >> 2d$

10. Total volume of atoms present in a face-centered cubic unit cell is

(a) $\frac{12}{3}\,\pi\, r^3$

(b) $\frac{16}{3}\,\pi\, r^3$

(c) $\frac{20}{3}\,\pi\, r^3$

(d) $\frac{24}{3}\,\pi\, r^3$

11. The number of Na and Mg atoms in the unit cell of bcc and fcc crystals, respectively, is

(a) 4 and 2
(b) 9 and 14
(c) 4 and 9
(d) 2 and 4

12. When A atoms occupy the corners of the unit cell and B atoms occupy the face centers in a fcc arrangement of A and B atoms, and one A atom is absent from one of the corners in the unit cell, simplest formula of the compound would be

(a) A_7B_3
(b) AB_3
(c) A_7B_{24}
(d) $A_{7/8}B_2$

13. The number of unit cell in 58.5 g of NaCl is approximately

(a) 6×10^{20}
(b) 1.5×10^{23}
(c) 6×10^{23}
(d) 0.5×10^{24}

14. 6:6 of NaCl coordination changes to 8:8 coordination on

(a) applying high pressure
(b) increasing temperature
(c) both
(d) remains unaffected on temperature and pressure change on coordi-| nation.

15. The type of fcc structure containing alternate tetrahedral void is

(a) CaF_2
(b) Na_2O
(c) ZnS
(d) NaCl

CHAPTER 4: PROBLEMS

1. What point groups are present in hexagonal crystal structure?
2. What do you mean by 222, mm2, and mmm point groups?
3. What do you understand by the 4_2 axis?
4. Explain 2_1 axis, mirror planes, and glide planes.
5. What point groups result on addition of two mirror planes at 90° to each other, (b) 60°?
6. Explain screw axis and center of symmetry.
7. Describe point groups and Bravais lattice.
8. Write the graphical symbols of rotation, inversion, axes, and mirror plane.
9. Explain 2, m, and $\frac{2}{m}$ point groups in a monoclinic crystal system.
10. Give the graphical representation of triclinic point group T in triclinic crystal system.

CHAPTER 5: PROBLEMS

Q.1 Why does a superconductor show infinite conductivity?

Q.2 Why does a superconductor exhibit diamagnetism?

Q.3 Explain K_3C_{60} acts as a superconductor.

Q4 Why does $(TMTSF)_2X$-type compounds behave as a superconductor?

Q.5 Explain high temperature superconductors (HTSC).

Q.6 Explain the mixed-oxide superconductor with example.

Q.7 If La^{3+} in La_2CuO_4 is replaced by Ba^{2+}, it becomes superconducting. Explain in detail.

Q.8 Explain the energy gap in a superconducting state.

Q.9 Explain why does $RbCs_2C_{60}$ behave as a superconductor?

CHAPTER 6: QUESTIONS

1. Explain the concept of phonons.
2. Explain phonon dispersion in one-dimensional monoatomic lattice and find the value of w and plot a graph between W and K.
3. What do you mean by Brillouin zone?
4. Explain why group velocity is zero at the edge of the zone?
5. Explain quantization of elastic waves as well as phonon momentum.
6. Explain Einstein's theory of heat capacity of solids.
7. Derive Debye's T^3 law. How does it explain the experimental results at low temperature? Show the graph between heat capacity in $J.mol^{-1}K^{-1}$ and $\frac{T}{\theta}$

CHAPTER 8: QUESTIONS

1. Explain the hysteresis loop of a ferro electric substance.
2. Explain the differences between a ferroelectric, pyroelectric, and piezoelectric substance.
3. Define the terms
4. Ionic polorizability (b) Dipolar polarizability (c) Electronic polarizability
5. Explain non-centro symmetric property of crystals.
6. Explain metal–metal and metal–semiconductor junction.
7. Describe Hall effect.
8. What do you mean by photovoltaic effect?
9. Explain the properties of dielectric materials.

CHAPTER 9: QUESTIONS

1. Distinguish between diamagnetic, paramagnetic, and ferromagnetic materials.
2. Discuss for the magnetic behavior of the following spinels (a) $ZnFe_2O_4$ (b) $MnFe_2O_4$ (c) $MgFe_2O_4$
3. Derive Langevin equation for diamagnetic substances.
4. Discuss the quenching of orbital contribution applying quantum mechanics.
6. Explain magnetization vector and permeability.

7. Explain Curie and Curie-Weiss law.
8. Explain Neel temperature.
9. Show the difference between a normal spinel and an inverse spinel.

CHAPTER 11: EXAMINATION PROBLEMS

Q.1 Define the following terms
(a) Molecular solid
(b) Covalent network solid
(c) Crystalline solid
(d) Isotropy
(e) Crystal lattice
(f) Cubic unit cells
(g) Face-centered cubic unit cell
(h) Ionic solid
(i) Metallic solid amorphous solid
(j) Anisotropy
(k) Bragg's equation
(l) Structure factor
(m) Schottky defects
(n) Frenkel defects
(o) Lattice energy
(p) X-ray diffraction
(q) Neutron diffraction
(r) Miller indices

Q.2 (a) Describe the theory of Bragg's method of crystal analysis.
(b) Differentiate between the cubic close packing and hexagonal close packing of spheres.

Q.3 (a) Derive the relation $n\lambda = 2d \sin \theta$ in crystallography.
(b) Draw diagrams to represent (i) F.C.C. lattice, (ii) B.C.C. lattice

Q.4 (a) Write a short note on Bravais Lattices.
(b) What do you understand by the packing efficiency of a crystal?
(c) Explain point groups and space groups.

Q.5 (a) Derive a relationship between the interplanar spacing of a crystal and the wavelength of X-ray diffracted by it.
(b) Differentiate between isomorphism and polymorphism.
(c) The face-centered cubic lattice has closer packing than body-centered cubic lattice. Why?

Q.6 (a) Explain the following with examples.
(i) Primitive unit cell and non-primitive unit cell.
(ii) Plane of symmetry and axis of symmetry.
(b) Define and explain the law of constancy of interfacial angles and the law of rational indices.

Q.7 (a) Discuss various elements of symmetry of a cubic crystal lattice.
(b) What do you understand by axis of four-fold symmetry and axis of three-fold symmetry?

(c) How many such axes are present in cubic crystals?

Q.8 (a) Write a note on point defects in ionic crystal.

(b) Discuss X-ray diffraction in elucidating structures of crystals and powders.

Q.9 (a) Explain the difference between TGA, DTA, and DSC.

(b) Explain the first, second, and third weight loss in TGA diagram in montmorillonite minerals.

(c) Explain the TGA for kinetic study.

(d) How can DTA be used for determination of phase transition?

Q.10 (a) Explain the intensity in X-ray diffraction of clay minerals.

(b) Explain glancing angle, d-value and planes in the PXRD of a crystal.

(c) Why first-order diffraction is considered in the characterization of solids using Bragg's equation?

(d) Does alkali chloride structure possess center of symmetry? Diffraction from (100) plane.

(e) Explain why X-ray powder diffraction pattern is used as a fingerprint for its identification.

Q.11 How does IR spectroscopy help in characterization of montmorillonite minerals?

Q.12 How does SEM analysis determine the extent of adsorption on solid surface?

Q.13 What symmetry elements do the following molecules possess

(a) CH_4 (b) NH_3 (c) CH_3Cl (d) BF_3 (e) C_6H_6 (f) C_2H_6 (g) H_2O (h) CO_2

Q.14 Copper crystal has fcc cubic lattice structure. Its density is 8.93 g cm^{-3}. What is the length of the unit cell? (N_A= 6.023 x 10^{23}; Atomic mass of Cu 63.5 g/ mol)?

Answer- 3.614 Å

Q.15 A body-centered cubic element of density 10.3 g.cm^{-3} has a cell edge of 314 pm. Calculate the atomic mass of element. (N_A= 6.023 x 10^{23}).

Answer - 94 amu

Q.16 Find the angles between the planes.

(a) (100) & (010) (b) (110) & (001)

(c) (100) & (001) (d) (110) & (111)

Solved Problems

Topic – Atoms, molecular formula calculation & Symmetry in crystals

Q.1 In a face-centered arrangement of A and B atoms, A atoms are at all the corners of the unit cell and B atoms are at all the face centers. For each unit cell, one A atom is missing from a corner position and one B atom is missing from one face position. The simplest formula of the resulting compound will be **[TIFR 2016]**

(a) $A_{14}B_{40}$
(b) A_7B_{20}
(c) $A_{1-x}B_{3-x}$
(d) AB_2

Answer – (b)

Solution:

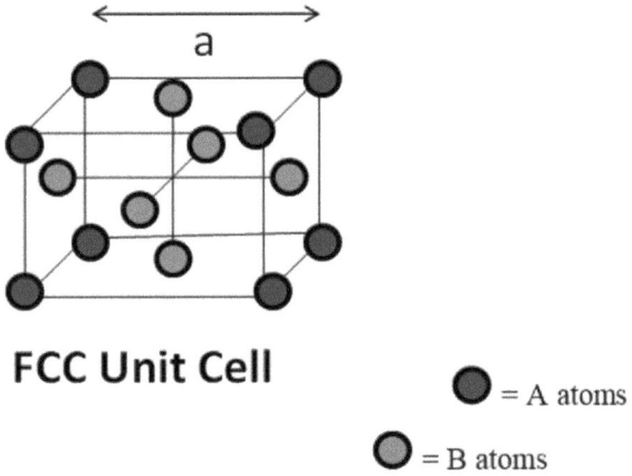

In a FCC unit cell, lattice points are at all the corners and face centers of a unit cell

Number of atoms in FCC unit cell = Z

$Z = 8(\text{corners}) \times \frac{1}{8}$ (contribution of corners) + $6(\text{face centers}) \times \frac{1}{2}$ (contribution of face)

$$Z = 4 \text{ (For FCC Unit Cell)}$$

In a given question, A atoms are at all the corners

B atoms are at all the face centers,

After missing one A atom from corners,

$$\text{Number of A atoms remain} = 7 \times \frac{1}{8} = \frac{7}{8}$$

After missing one B atom from face centers,

$$\text{Number of B atoms remain} = 5 \times \frac{1}{2} = \frac{5}{2}$$

Thus, A:B:: $\frac{7}{8} : \frac{5}{2}$ taking simplest ratio by multiply through 8 and we get,

A:B = 7:20, Hence the simplest formula is A_7B_{20}

Answer – option (b) is correct.

Q.2 How many atoms are there in an element packed in a fcc structure

[NET JUNE 2012]

(a) 1
(b) 2
(c) 4
(d) 8

Answer – (c)
Solution:

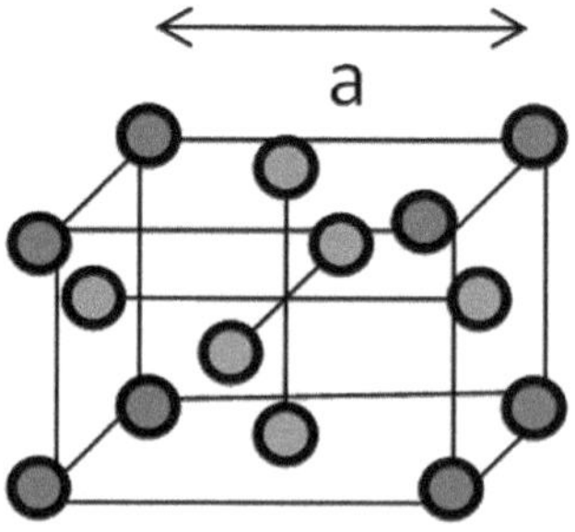

FCC Unit

In a FCC unit cell, lattice points are at all the corners and face centers of a unit cell

Number of atoms in FCC unit cell = Z

$$Z = 8(\text{corners}) \times \frac{1}{8} \text{ (contribution of corners)} + 6(\text{face centers}) \times \frac{1}{2}$$
(contribution of face)

$$Z = 4 \text{ (For FCC Unit Cell)}$$

Answer – option (c) is correct.

Q.3 A compound of M and X atoms has a cubic unit cell. M atoms are at the comers and body-center position and X atoms are at face-center positions of the cube. The molecular formula of the compound is

[NET DEC. 2013]

(a) MX
(b) MX_2
(c) M_3X_2
(d) M_2X_3

Answer – (d)
Solution:

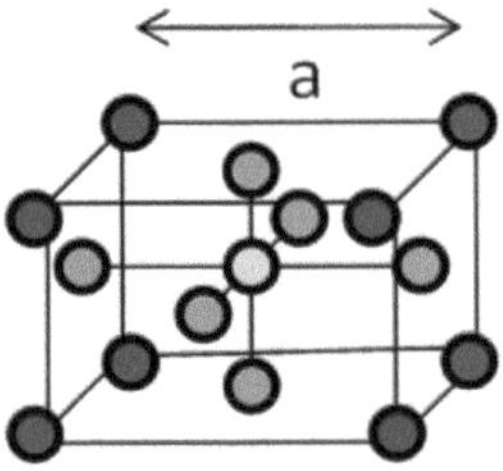

Cubic Unit

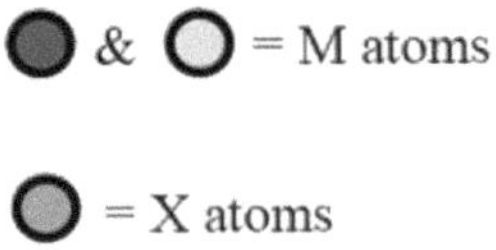

In a given question, M atoms are at all the corners and body center and X atoms are at all the face centers of a cubic unit cell

Total M atoms in cubic unit cell = $8 \times \frac{1}{8} + 1 \times 1$(contribution of body center)

Total M atoms = 2

$$\text{Total X atoms in cubic unit cell} = 6 \text{ x } \frac{1}{2} = 3$$

Thus, M:X:: 2:3,

Hence, simplest formula of compound is M_2X_3
Answer – option (d) is correct.

Q.4 The lattice parameters for a monoclinic crystal are **[GATE 2008]**

(a) $a \neq b \neq c; \alpha = \gamma = 90°$
(b) $a = b \neq c; \alpha = \beta \neq \gamma$
(c) $a \neq b \neq c; \alpha \neq \beta \neq \gamma$
(d) $a = b = c; \alpha = \gamma = 90°$

Answer – (a)
Solution:

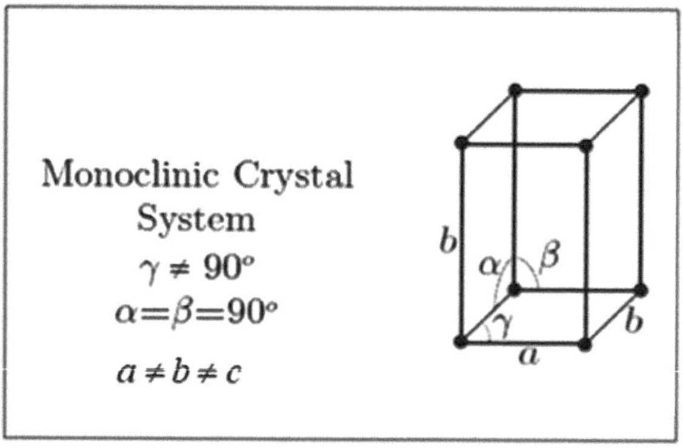

The lattice parameters for a monoclinic crystal are clearly shown in the above diagram

Hence, $a \neq b \neq c; \alpha = \gamma = 90°$, these parameters belong from monoclinic crystal

Answer – option (a) is correct.

Topic–Packing fraction density, coordinate system & relation between edge length, radius and nearest neighbor distance

Q.5 The lattice parameter of an element stabilized in a fcc structure is 4.04 Å. The atomic radius of the element is **[NET JUNE 2012]**

(a) 2.86 Å

(b) 1.43 Å
(c) 4.29 Å
(d) 5.72 Å

Answer – (b)
Solution:

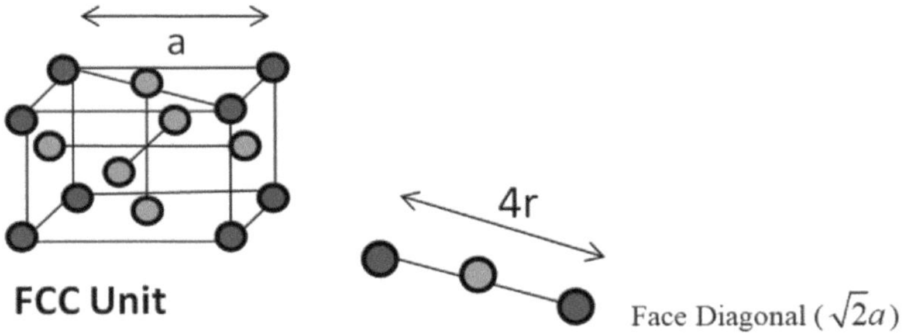

In a given question, lattice parameter (edge length) of FCC structure (a) = 4.04 Å

The relation between radius of element and lattice parameter in FCC is ($4r = \sqrt{2}a$)

$$r = \frac{\sqrt{2}}{4} \times a,\ r = \frac{\sqrt{2}}{4} \times 4.04\ \text{Å}$$

$$r = 1.414 \times 1.01$$

$$r = 1.43\ \text{Å}$$

Answer – option (b) is correct.

Q.6 A metal crystallizes in fcc structure with a unit cell side of 500 pm. If the density of the crystal is 1.33 g/cm^3, the molar mass of the metal is close to **[NET DEC. 2012]**

(a) 23
(b) 24
(c) 25
(d) 26

Answer – (c)

Solution: density of crystal $\rho = \frac{ZM_w}{N_A a^3}$,

where, ρ = density of crystal, a^3 = volume of cubic crystal

Z = Number of atoms in Unit cell, For FCC structure (Z) = 4

M_w = Molar mass of metal, N_A = Avogadro's number, $N_A = 6.023 \times 10^{23}$ mol^{-1}

In a given question, $\rho = 1.33\ g/cm^3$, $M_w = 500$ pm, $M_w = ?$

From density formula, $M_w = \dfrac{\rho \times N_A \times a^3}{Z}$

$$M_w = \frac{1.33 \times 6.023 \times 10^{23} \times \left(500 \times 10^{-10}\right)^3}{4}, \ (1\ pm = 10^{-10}\ cm)$$

$$M_w = 25.03\ g\ /mol^{-1} \text{ or close to } 25$$

Answer – option (c) is correct.

Q.7 The packing fraction of a simple cubic lattice is close to **[NET DEC. 2012]**

(a) 0.94
(b) 0.76
(c) 0.52
(d) 0.45

Answer – (c)
Solution:

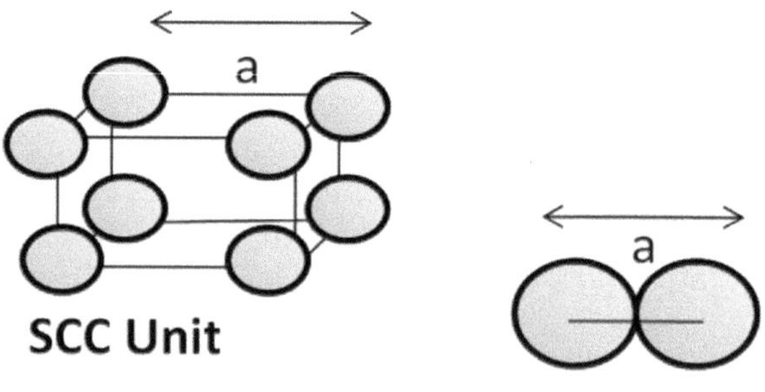

In a SCC lattice, atoms are located only at the corners of the cubic unit cell (Z = 1) the relation between radius of particle (r) and lattice parameter (a) in SCC is

$$a = 2r$$

$$\text{the volume of cubic unit cell (V)} = a^3 = (2r)^3 = 8r^3$$

$$\text{the volume of occupied space of particle(atom)} = \frac{4}{3}\pi r^3$$

Packing fraction or packing efficiency for SCC

$$= \frac{Volume\ of\ one\ atom}{Volume\ of\ cubic\ unit\ cell(V)}$$

$$\text{Packing fraction} = \frac{1(Z) \times \frac{4}{3}\pi r^3}{a^3}$$

$$\text{Packing fraction} = \frac{4\pi r^3}{3 \times 8r^3}, \quad a^3 = 8r^3$$

$$\text{Packing fraction} = \frac{\pi}{6} = \frac{3.14}{6} = 0.52$$

Answer – option (c) is correct.

Q.8 A metal crystallizes in the face-centered cubic lattice parameter of 4.20 Å. The shortest atom to atom contact distance in the lattice is **[GATE 2013]**

(a) 4.20 Å
(b) 2.97 Å
(c) 2.42 Å
(d) 2.10 Å

Answer – (b)
Solution:

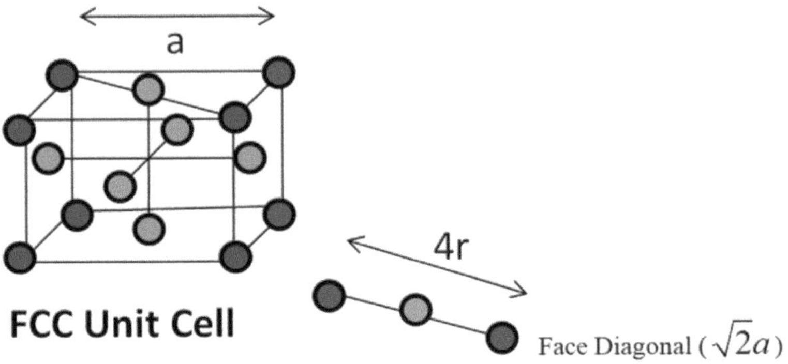

In given question, lattice parameter (a) = 4.20 Å

The relation between radius of atom (r) and lattice parameter in FCC is ($4r = \sqrt{2}a$)

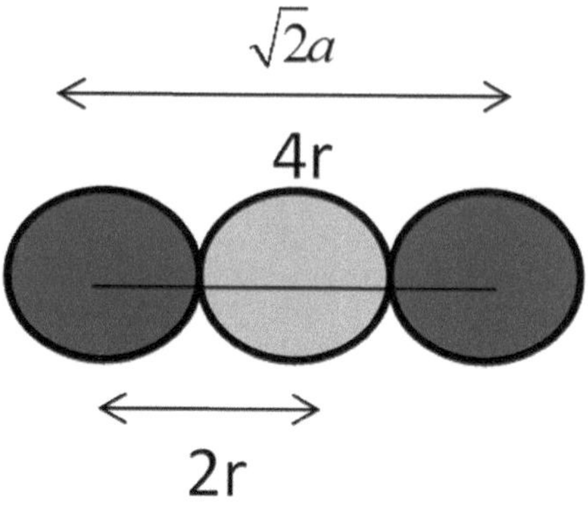

The shortest atom to atom contact distance is (2r)

$$4r = \sqrt{2}a$$

$$2r = \frac{\sqrt{2}a}{2},\ 2r = \frac{\sqrt{2} \times 4.20}{2}$$

Shortest atom to atom contact distance ($2r = 2.97$ Å)
Answer – option (b) is correct.

Q.9 The coordinates for the atoms in a body-centered cubic unit cell are **[NET JUNE 2013]**

(a) (0, 0, 0) and (1/2, 0, 0)
(b) (0, 0, 0) and (1/2, 1/2, 1/2)
(c) (0, 0, 0) and (0, 1/2, 0)
(d) (0, 0, 0) and (0, 0, 1/2)

Answer – (b)
Solution:

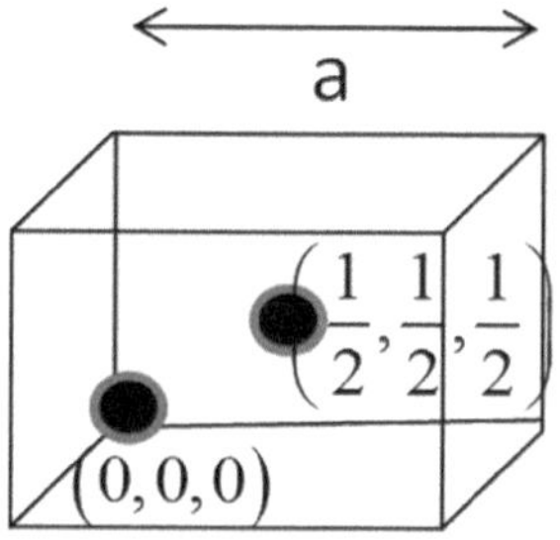

BCC Unit Cell

In BCC lattice, atoms present at all the corners and body center of a cubic unit cell

$Z = 8$ (corners) x $\frac{1}{8}$ (contribution of corners) + 1(body center) x 1(contribution of body center)

Total atoms in BCC lattice (Z) = 2

So, one atom is present at the origin point of cubic unit cell whose coordinates are

(0, 0, 0) and the other atom is present at the body center of cubic unit cell whose coordinates from origin are (1/2, 1/2, 1/2).

Answer – option (b) is correct.

Q.10 Molybdenum crystallizes in a bcc structure with unit cell dimensions of 0.314 nm. Considering the atomic mass of molybdenum to be 96 g/mol, its density (in kg/m^3) is -------------------**[GATE 2014]**

Answer – 10000–10500

Solution:

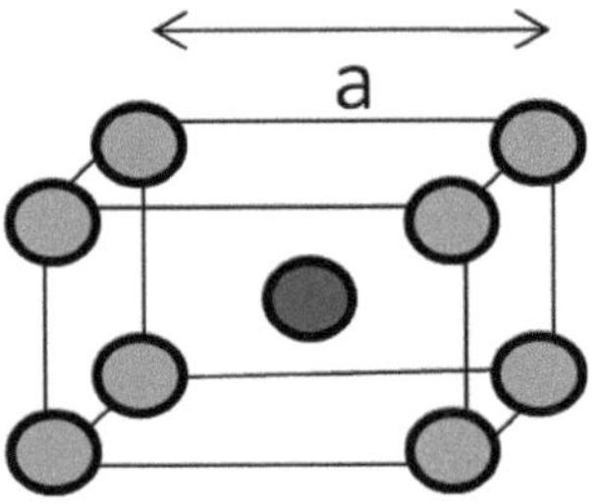

BCC Unit

In a given question, cell dimension or edge length (a) = 0.314 nm or 3.14×10^{-10} m

Atomic mass of molybdenum (M_w) = 96 g/mol or 96×10^{-3} kg/mol,

ρ (density) = ?

For BCC structure, Total atoms in BCC lattice (Z) = 2

Density of crystal is calculated as $\rho = \frac{ZM_w}{N_A a^3}$,

where ρ = density of crystal, a^3 = volume of cubic crystal

M_w = Molar mass of metal, N_A = Avogadro's number, $N_A = 6.023 \times 10^{23}$ mol^{-1}

Now, $$\rho = \frac{2 \times 96 \times 10^{-3}}{6.023 \times 10^{23} \times \left(3.14 \times 10^{-10}\right)^3}$$

$$\rho = 10297.7 \text{ kg/m}^3$$

Answer – the calculated value of density is lies between range of 10000–10500.

Q.11 The packing factor (PF) and number of atomic sites per unit cell (N) of an FCC crystal system are **[NET JUNE 2015]**

(a) P F = 0.52 and N = 3
(b) PF = 0.74 and N = 3
(c) PF = 0.52 and N = 4
(d) PF = 0.74 and N = 4

Answer – (d)
Solution:

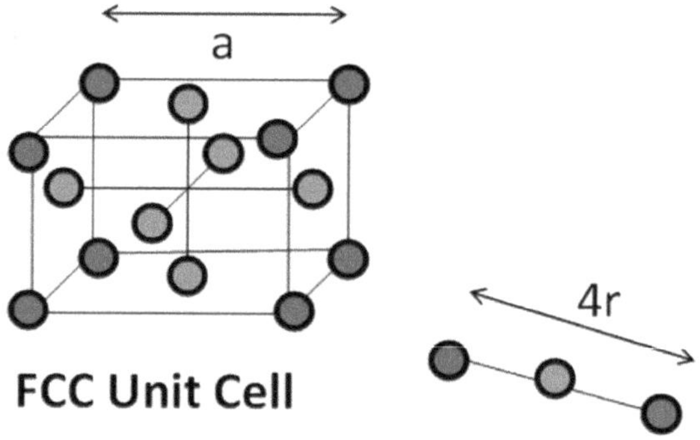

In an FCC unit cell, lattice points are at all the corners and face centers of unit cell

Number of atomic sites in FCC unit cell = N

N = 8 (corners) x $\frac{1}{8}$ (contribution of corners) + 6(face centers) x $\frac{1}{2}$ (contribution of face)

N = 4 (for FCC)

the relation between radius of particle (r) and lattice parameter (a) in FCC is

$$4r = \sqrt{2}a, \quad a = 2\sqrt{2}r$$

the volume of cubic unit cell (V) = $a^3 = (2\sqrt{2}r)^3 = 16\sqrt{2}r^3$

the volume of occupied space of particle (atom) = $\frac{4}{3}\pi r^3$

packing factor or packing efficiency for FCC = $\frac{Volume\ of\ four\ atom}{Volume\ of\ cubic\ unit\ cell(V)}$

$$\text{Packing factor (PF)} = \frac{4(N)\times\frac{4}{3}\pi r^3}{a^3}$$

$$\text{Packing factor (PF)} = \frac{16\pi r^3}{3\times 16\sqrt{2}r^3}$$

$$\text{Packing factor (PF)} = \frac{\pi}{3\sqrt{2}} = \frac{3.14}{3\times 1.414} = 0.74$$

$$\text{PF} = 0.74, \text{N} = 4$$

Answer – option (d) is correct.

Q.12 Both NaCl and KCI crystallize with the FCC structure. However, the X-ray powder diffraction pattern of whereas that KCl corresponds NaCl corresponds to the FCC structure whereas that of KCl corresponds to simple cubic structure. This is because **[NET DEC. 2013]**

(a) K+ and Cl- are isoelectronic
(b) Na^+ and Cl^- are isoelectronic
(c) K^+ and Cl^- are disordered in the crystal lattice
(d) KCI has anti-site defects

Answer – (a)
Solution: Since K^+ and Cl^- are isoelectronic species, some of the peaks are missing completely. The scattering intensities for X-rays are directly related to the number of electrons in the atoms. Light atoms scatter X-rays weakly. While the heavy atoms scatter X-rays more effectively.

Q.13 If the atoms/ions in the crystal are taken to be hard spheres touching each other in the unit cell, then the fraction of volume occupied in the body-centered cubic structure is **[NET DEC. 2017]**

(a) $\sqrt{3}\pi$

(b) $\frac{\sqrt{2}\pi}{6}$

(c) $\frac{\pi}{6}$

(d) $\frac{\sqrt{3}\pi}{8}$

Answer – (d)
Solution:

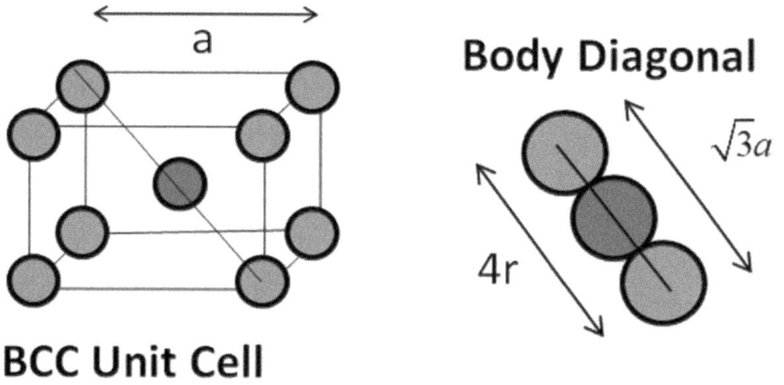

In BCC lattice, atoms present at all the corners and body center of a cubic unit cell

Z = 8 (corners) x $\frac{1}{8}$(contribution of corners) + 1(body center) x 1 (contribution of body center)

Total atoms in BCC lattice (Z) = 2

The relation between radius of atom (r) and lattice parameter in FCC is ($4r = \sqrt{3}a$)

From relation, $a = \frac{4}{\sqrt{3}}r$, Volume of Cubic unit cell (V) = $a^3 = \left(\frac{4}{\sqrt{3}}r\right)^3$, the volume of occupied space of particle (atom) = $\frac{4}{3}\pi r^3$

Packing fraction or packing efficiency for BCC

$$= \frac{Volume\ of\ two\ atom}{Volume\ of\ cubic\ unit\ cell(V)}$$

$$\text{Packing fraction} = \frac{2(Z) \times \frac{4}{3}\pi r^3}{a^3}$$

$$\text{Packing fraction} = \frac{8\pi r^3}{3\times\frac{64}{3\sqrt{3}}r^3}, \qquad a^3 = \left(\frac{4}{\sqrt{3}}r\right)^3$$

$$\text{Packing fraction} = \frac{\sqrt{3}\pi}{8} \text{ or } 0.68$$

Answer – option (d) is correct.

Q.14 The (002) plane of an elemental FCC crystal diffracts X-rays (λ = 0.154 nm) at Bragg angle 90°. The density of the crystal is 4 x 10^4 kg m^{-3}. The atomic weight of the elemental solid is **[NET JUNE 2018]**

(a) 22
(b) 44
(c) 88
(d) 66

Answer – (a)

Solution: In a given question, Bragg's angle (θ) = 90°, density (ρ) = 4 x 10^4 kg m^{-3}

X-ray wavelength (λ) = 0.154 nm, plane (*hkl*) = (002), M_w = ?

For FCC crystal (Z) = 4

$$\text{Interplanar spacing } (d_{hkl}) = \frac{a}{\sqrt{h^2+\kappa^2+l^2}} = \frac{a}{\sqrt{0^2+0^2+2^2}} = \frac{a}{2}$$

Use Bragg's equation to calculate lattice parameter (a)

Bragg's equation = $2d_{hkl}\sin\theta = n\lambda$, where n = order of reflection

$n = 1$ (if not given in question)

$2d_{hkl}\sin\theta = n\lambda$, $2\times\frac{a}{2}\times\sin 90^o = 1\times 0.154nm$, a = 0.154 nm or 1.54 x 10^{-10} m

Now use density formula to calculate atomic mass

$$\text{Density of crystal } (\rho) = \frac{ZM_w}{N_A a^3}$$

$$M_w = \frac{\rho\times N_A\times a^3}{Z}, \quad M_w = \frac{4\times10^4\times6.023\times10^{23}\times\left(1.54\times10^{-10}\right)^3}{4}$$

$$M_w = 0.022 \text{ kg/mol or } = 22 \text{ g/mol}$$

Answer – option (a) is correct.

Q.15 Consider that AgX crystallizes in rock salt structure. The density of AgX is 6477 kg/m^3 and unit cell length is 577.5 pm. Atomic weight of Ag is 107.87 g mol^{-1}. The atomic weight of X (in g mol^{-1}, rounded off to two decimal places) is------------------------- **[GATE 2020]**

Answer – 79.97

Solution: In a given question, density (ρ) = 6477 kg/m^3, cell length (a) = 577.5 pm

Atomic weight of Ag = 107.87 g/mol, atomic weight of X = ?

AgX Crystal = Rock salt structure = NaCl Structure = FCC lattice, Z = 4

$$\text{Density of crystal } (\rho) = \frac{ZM_w}{N_A a^3}$$

$$M_w = \frac{\rho \times N_A \times a^3}{Z},\ M_w = \frac{6477 \times 6.023 \times 10^{23} \times \left(577.5 \times 10^{-12}\right)^3}{4}$$

$M_w = 0.1878$ kg/mol or 187.8 g/mol of AgX Crystal in rock salt structure

AgX (M_w) = atomic weight of Ag + atomic weight of X

Atomic weight (X) = 187.8 – 107.87 = 79.97 g/mol

Answer – atomic weight of X is 79.97 g/mol.

Topic – Miller indices, Bragg's equations, interplanar spacing & Crystal defects

Q.16 For a face-centered cubic lattice, the miller indices for the first Bragg's peak (smallest Bragg angle) are **[GATE 2012]**

(a) 002
(b) 111
(c) 001
(d) 110

Answer – (b)

Solution: Condition for allowed reflection for FCC lattice is that all the miller indices (*hkl*) are either even or odd.

Option (c) and (d) are wrong because both options did not fulfill the allowed reflection condition for FCC lattice. But option (a) and (b) both fulfilled the condition of allowed reflection for FCC lattice.

Now, option (a) hkl = 002, $d_{hkl} = \frac{a}{\sqrt{h^2+k^2+l^2}}, d_{002} = \frac{a}{\sqrt{0^2+0^2+2^2}}$ $= \frac{a}{2} = 0.5a$

Option (b) $hkl = 111$, $d_{hkl} = \frac{a}{\sqrt{h^2+k^2+l^2}}, d_{111} = \frac{a}{\sqrt{1^2+1^2+1^2}} = \frac{a}{\sqrt{3}} = 0.58a$

From Bragg's equation, $2d_{hkl} \sin\theta = n\lambda$

$\sin\theta = \frac{n\lambda}{2d_{hkl}}$, angle ($\theta$) is inversely related with d_{hkl} and relation clearly shows that more the value of interplanar spacing (d_{hkl}), less or small will be Bragg's angle.

Hence, miller indices (111) show the smallest angle among all the given option.

Answer – option (b) is correct.

Q.17 The angle between the two planes represented by the Miller indices (110) and (111) in a simple cubic lattice **[NET DEC. 2012]**

(a) 30°
(b) 45°
(c) 60°
(d) None of these

Answer (d)

Solution: In a given question, miller indices are (110) and (111). The angle between two planes is calculated as $\cos\theta = \frac{h_1h_2 + k_1k_2 + l_1l_2}{\sqrt{h_1^2+k_1^2+l_1^2}\sqrt{h_2^2+k_2^2+l_2^2}}$, where $h_1k_1l_1 = (110)$, $h_2k_2l_2$ (111)

$$\cos\theta = \frac{1\times1+1\times1+0\times1}{\sqrt{1^2+1^2+0^2}\sqrt{1^2+1^2+1^2}}$$

$$\cos\theta = \sqrt{\frac{2}{3}}, \theta = \cos^{-1}\sqrt{\frac{2}{3}} = 35.7^o$$

Answer – option (d) is correct.

Q.18 Crystal A diffracts from (111) and (200) planes but not from (110) plane, while the crystal B diffracts from (110) and (200) planes but not from the (111) plane. From the above, we may conclude that **[NET DEC. 2012]**

(a) A has FCC lattice while B has BCC lattice
(b) A has BCC lattice while B has FCC lattice
(c) A and B both have FCC lattice
(d) A and B both have BCC lattice

Answer – (a)
Solution: Selection rule for allowed reflection = structure factor (S) should be non-zero.

Selection rules in terms of miller indices are given below.

For BCC Lattice $h + k + l$ = even = Allowed reflection
$h + k + l$ = odd = Forbidden reflection

For FCC Lattice (h,k,l) are either even or odd = Allowed reflection
(h,k,l) are mix (even or odd) = Forbidden reflection

In a given question, crystal A diffracts from (111) & (200) but not from (110) plane. It is clearly that the selection rule of miller indices for FCC lattice reflect only either even or odd plane but not reflect mix (even or odd) plane. So, Crystal A is FCC lattice.

While crystal B diffracts from (110) & (200) but not from (111) plane. This is because of selection rule of allowed reflection for BCC structure

For Allowed reflection = $h + k + l$ = even, 1+ 1+0 = Even, 2+0+0 = even,

But for plane (111), $h + k + l$ = odd, 1+1+1 = odd, for Forbidden reflection.

Hence, Crystal B is BCC lattice.

Answer – option (a) is correct.

Q.19 A plane of spacing "d" shows first-order Bragg diffraction at angle θ. A plane of spacing "2d" **[NET JUNE 2013]**

(a) Shows Bragg diffraction at 2θ

(b) Shows Bragg diffraction at $\frac{\theta}{2}$

(c) Shows Bragg diffraction at $\sin^{-1}\left(\frac{\sin\theta}{2}\right)$

(d) Shows Bragg diffraction at $\sin^{-1}\left(\frac{\sin 2\theta}{2}\right)$

Answer – (c)

Solution: Bragg's equation = $2d\sin\theta = n\lambda$, where n = order of reflection

$n = 1$ (first-order Bragg's diffraction), λ = wavelength,
d = interplanar spacing

Use Bragg's equation: $2d \sin\theta = 1 \times \lambda, \lambda = 2d \sin\theta$

Now use interplanar spacing "2d" in Bragg's equation and we get

$$2 \times (2d) \sin\theta = 1 \times \lambda$$

$$\sin\theta = \frac{\lambda}{4d}, \quad \sin\theta = \frac{2d\sin\theta}{4d}, \because \lambda = 2d\sin\theta$$

$$\sin\theta = \frac{\sin\theta}{2}, \theta = \sin^{-1}\left(\frac{\sin\theta}{2}\right)$$

Answer – option (c) is correct.

Q.20 Silver crystallizes in face-centered cubic structure. The 2nd-order diffraction angle of a beam of X-ray (λ = 1Å) of (111) plane of the crystal is 30°. Therefore, the unit cell length of the crystal would be **[NET JUNE 2015]**

(a) a = 3.151 Å

(b) a = 3.273 Å

(c) a = 3.034 Å

(d) a = 3.464 Å

Answer – (d)

Solution: In a given question, wavelength (λ)= 1 Å or 1 x 10^{-10} m, (hkl) = (111),

n = 2nd-order, Bragg's angle (θ) = 30°, cell length (a) = ?

For plane (111), interplanar distance $d_{hkl} = \frac{a}{\sqrt{h^2 + k^2 + l^2}}$,

$$d_{111} = \frac{a}{\sqrt{1^2 + 1^2 + 1^2}} = \frac{a}{\sqrt{3}}$$

Use Bragg's equation, $2d_{hkl} \sin\theta = n\lambda$, $2 \times \frac{a}{\sqrt{3}} \sin 30^o = 2 \times 1, \quad \because \quad \lambda = 1\text{Å}$

Hence, Cell length $(a) = 2\sqrt{3} = 3.464\,\text{Å} \quad \because \sin 30^o = \frac{1}{2}$

Answer – option (d) is correct.

Q.21 For a simple cubic crystal, X-ray diffraction shows intense reflections for angles θ_1 and θ_2 which are assigned sin to [101] and [111] planes, respectively, the ratio $\frac{\sin\theta_1}{\sin\theta_2}$ is **[NET DEC. 2015]**

(a) 1.5
(b) 1.22
(c) 0.82
(d) 0.67

Answer – (c)

Solution: In a given question, $(h_1k_1l_1) = (101)$ and $(h_2k_2l_2) = (111)$

Now, interplanar spacing $(d_{h_1k_1l_1}) = \frac{a}{\sqrt{h^2+k^2+l^2}}, d_{101} = \frac{a}{\sqrt{1^2+0^2+1^2}} = \frac{a}{\sqrt{2}}$

$$(d_{h_2k_2l_2}) = \frac{a}{\sqrt{h^2+k^2+l^2}}, d_{111} = \frac{a}{\sqrt{1^2+1^2+1^2}} = \frac{a}{\sqrt{3}}$$

Now using Bragg's equation, $2d_{hkl}\sin\theta = n\lambda$,

$$2d_{h_1k_1l_1}\sin\theta_1 = n\lambda \qquad \text{----------Eq.(1)}$$

$$2d_{h_2k_2l_2}\sin\theta_2 = n\lambda \qquad \text{----------Eq.(2)}$$

Eq.(1)/Eq.(2) and we get, $\frac{2}{2}\times\frac{\sin\theta_1}{\sin\theta_2} = \frac{d_{101}}{d_{111}}\times\frac{n\lambda}{n\lambda} = \frac{\frac{a}{\sqrt{3}}}{\frac{a}{\sqrt{2}}}$

Hence, $\frac{\sin\theta_1}{\sin\theta_2} = \sqrt{\frac{2}{3}} = \frac{1.414}{1.732} = 0.82$

Answer – option (c) is correct.

Q.22 The angle at which the first-order Bragg reflection is observed from (110) plane in a simple cubic unit cell of side 3.238 Å, when chromium K_α radiation of wavelength 2.2 Å is used, is **[NET DEC. 2013]**

(a) 30°
(b) 45°
(c) 60°
(d) 90°

Answer – (a)

Solution: In a given question, plane (hkl) = (110), Cell length (a) = 3.238 Å Wavelength (λ) = 2.2 Å, n = 1 (first-order reflection), angle (θ) = ?

Interplanar spacing $(d_{hkl}) = \dfrac{a}{\sqrt{h^2+k^2+l^2}}, d_{110} = \dfrac{a}{\sqrt{1^2+1^2+0^2}} = \dfrac{a}{\sqrt{2}}$

Now using Bragg's equation, $2d_{hkl}\sin\theta = n\lambda$

$$2\times\frac{a}{\sqrt{2}}\sin\theta = 1\times 2.2,\quad \sin\theta = \frac{2.2\times\sqrt{2}}{2\times 3.238} \quad \because \quad (a) = 3.238\ \text{Å}$$

$$\sin\theta = 0.5 = \frac{1}{2},\quad \theta = \sin^{-1}\left(\frac{1}{2}\right) = 30^o$$

Answer – option (a) is correct.

Q.23 An element exists in two crystallographic modifications with FCC and BCC structures. The ratio of the densities of the FCC and BCC modifications in terms of the volumes of their unit cells (V_{FCC} and V_{BCC}) is **[NET DEC. 2013]**

(a) $V_{BCC} : V_{FCC}$

(b) $2V_{BCC} : V_{FCC}$

(c) $V_{BCC} : 2V_{FCC}$

(d) $V_{BCC} : \sqrt{2}\ V_{FCC}$

Answer – (b)

Solution: For FCC lattice $(Z_{FCC}) = 4$, where Z is number of atoms in a unit cell

For BCC lattice $(Z_{BCC}) = 2$

Now, density of crystal is calculated as

$$(\rho) = \frac{ZM_w}{N_A V} = \frac{ZM_w}{N_A a^3} \quad \because \quad (V = a^3)$$

Now,

$$\rho_{FCC} = \frac{Z_{FCC}M_w}{N_A V_{FCC}} = \frac{4\times M_w}{N_A V_{FCC}} \quad \text{-------------Eq. (1)}$$

$$\rho_{BCC} = \frac{Z_{BCC}M_w}{N_A V_{BCC}} = \frac{2\times M_w}{N_A V_{BCC}} \quad \text{------------Eq. (2)}$$

Now, Eq. (1)/Eq. (2) and we get, $\dfrac{\rho_{FCC}}{\rho_{BCC}} = \dfrac{\dfrac{4\times M_w}{N_A V_{FCC}}}{\dfrac{2\times M_w}{N_A V_{BCC}}} = \dfrac{2V_{BCC}}{V_{FCC}}$

Hence, ratio of densities is $\rho_{FCC} : \rho_{BCC} :: 2V_{BCC} : V_{FCC}$

Answer – option (b) is correct.

Q.24 The $\sin^2 \theta$ values obtained from X-ray powder diffraction pattern of a solid are 2x, 4x, 6x, and 8x, where x is equal to 0.06. The wavelength of X-ray used to obtain this pattern is 1.54 Å. The unit cell and the unit cell length, respectively are **[NET DEC. 2016]**

(a) BCC, 3.146 Å
(b) FCC, 3.146 Å
(c) SCC, 6.281 Å
(d) BCC, 1.544 Å

Answer – (a)

Solution: In a given question, wavelength (λ) = 1.54 Å, $\sin^2 \theta$ obtained pattern are (2x, 4x, 6x, and 8x), where x = 0.06, $n = 1$ (suppose first-order reflection) cell length (a) = ?

Interplanar Spacing $(d_{hkl}) = \dfrac{a}{\sqrt{h^2+k^2+l^2}}$ on squaring both sides and we get $(d^2_{hkl}) = \dfrac{a^2}{h^2+k^2+l^2}$

Now use Bragg's equation ($2d_{hkl} \sin \theta = n\lambda$).

On squaring Bragg's equation both sides and we get

$$4d^2_{hkl} \sin^2 \theta = n^2\lambda^2, \ \sin^2 \theta = \frac{n^2\lambda^2}{4d^2_{hkl}}$$

$$\sin^2 \theta = \frac{n^2\lambda^2}{4a^2}\left(h^2+k^2+l^2\right) \quad \because \quad (d^2_{hkl}) = \frac{a^2}{h^2+k^2+l^2}$$

Now, $\dfrac{n^2\lambda^2}{4a^2}$ is a constant value and is equal to x, $\dfrac{n^2\lambda^2}{4a^2} = \text{x}$

$$a^2 = \frac{1^2 \times (1.54)^2}{4 \times 0.06} \quad \because \quad \text{x} = 0.06$$

Cell length (a) = 3.146 Å

Options (c) and (d) are wrong due to wrong value of cell length (a)
Now, check option (a) and (b)
Selection rule for allowed reflection for Miller indices (*hkl*)
For SCC lattice = any value of Miller indices (*hkl*)
For BCC lattice = ($h + k + l$) should be even

For FCC lattice = miller indices (*hkl*) either all even value or odd value give allowed reflection.

Now, compare modified Bragg's equation with obtained pattern (2x, 4x, 6x, 8x)

$$\sin^2\theta = \frac{n^2\lambda^2}{4a^2}\left(h^2+k^2+l^2\right),\quad \frac{n^2\lambda^2}{4a^2} \text{ x and } \left(h^2+k^2+l^2\right) = 2, 4, 6, 8$$

For (2x), 1. $(h^2+k^2+l^2) = (1^2+1^2+0^2) = (1+1+0)$ = even value = BCC lattice

For (4x), 2. $(h^2+k^2+l^2) = (2^2+2^2+0^2) = (2+0+0)$ = even value = BCC lattice

Answer – option (a) is correct.

Q.25 The correct relationship among the following, for a tetragonal ($a = b \neq c$; $\alpha = \beta = \gamma = 90°$) crystal system, is **[NET JUNE 2019]**

(a) $\sin^2\theta = \dfrac{\lambda^2}{4a^2}\left[c^2\left(h^2+k^2\right)+a^2l^2\right]$

(b) $\sin^2\theta = \dfrac{\lambda^2}{4a^2c^2}\left[c^2\left(h^2+k^2\right)+a^2l^2\right]$

(c) $\sin^2\theta = \dfrac{\lambda^2}{4c^2}\left[a^2\left(h^2+k^2\right)+c^2l^2\right]$

(d) $\sin^2\theta = \dfrac{\lambda^2}{4a^2}\left[h^2+k^2+l^2\right]$

Answer – (b)

Solution: In a given question, tetragonal crystal system ($a = b \neq c$; $\alpha = \beta = \gamma = 90° = 90°$)

Interplanar spacing (d_{hkl}) calculated as, $\dfrac{1}{d^2_{hkl}} = \left(\dfrac{h}{a}\right)^2 + \left(\dfrac{k}{b}\right)^2 + \left(\dfrac{l}{c}\right)^2$

$$\frac{1}{d^2_{hkl}} = \left(\frac{h}{a}\right)^2 + \left(\frac{k}{a}\right)^2 + \left(\frac{l}{c}\right)^2 \quad \because \quad (a = b \neq c)$$

$$\frac{1}{d^2_{hkl}} = \left(\frac{h^2+k^2}{a^2}\right) + \left(\frac{l^2}{c^2}\right),\quad d^2_{hkl} = \frac{a^2c^2}{c^2\left(h^2+k^2\right)+a^2l^2}$$

Now use Bragg's equation ($2d_{hkl}\sin\theta = n\lambda$),

On squaring Bragg's equation both sides and we get

$$4d^2_{hkl}\sin^2\theta = n^2\lambda^2,\quad \sin^2\theta = \frac{n^2\lambda^2}{4d^2_{hkl}},\quad n = 1 \text{ (if not given in question)}$$

$$\sin^2\theta = \frac{1^2\lambda^2}{4d^2_{hkl}}, \quad \sin^2\theta = \frac{\lambda^2}{4a^2c^2}[c^2\left(h^2+k^2\right)+a^2l^2], \because$$

$$d^2_{hkl} = \frac{a^2c^2}{c^2\left(h^2+k^2\right)+a^2l^2}$$

Answer – option (b) is correct.

Q.26 X-rays of 173 pm wavelength are reflected by the (111) plane of a cubic primitive crystal at 30°. The unit cell length (in pm) is closest to **[NET DEC. 2019]**

(a) 173
(b) 300
(c) 346
(d) 600

Answer – (b)
Solution: In a given question, Wavelength (λ) = 173 pm, Plane = (hkl) = (111)

Bragg's angle (θ) = 30°, $n = 1$ (if not given in question), Cell length (a) = ?

Use interplanar spacing $(d_{hkl}) = \frac{a}{\sqrt{h^2+k^2+l^2}}, d_{111} = \frac{a}{\sqrt{1^2+1^2+1^2}} = \frac{a}{\sqrt{3}}$

Now use Bragg's equation to calculate cell length (a)

$$2d_{hkl}\sin\theta = n\lambda, \quad 2\times\frac{a}{\sqrt{3}}\times\sin 30^o = 1\times 173 \quad \because d_{111} = \frac{a}{\sqrt{3}}$$

(a = 299.64) pm or close to 300 pm

Answer – option (b) is correct.

Practice Problems

Q.1 A compound AxBy has a cubic structure with A atoms occupying all comers of the cube as well as all the face-center positions. The B atoms occupy four tetrahedral voids. The values of x and y, respectively, are **[NET JUNE 2014]**

(a) 4,4
(b) 4,8
(c) 8,4
(d) 4,2

Q.2 Each void in a two-dimensional hexagonal close-packed layer of circles is surrounded by **[NET DEC. 2016]**

(a) Six circles
(b) Three circles.
(c) Four circles
(d) Twelve circles

Q.3 A crystal has the lattice parameters $a \neq b \neq c$ and $\alpha = \beta = \gamma = 90°$. The crystal system is **[GATE 2013]**

(a) Tetragonal
(b) Monoclinic
(c) Cubic
(d) Orthorhombic

Q.4 Metallic gold crystallizes in a face-centered cubic lattice with the lattice parameter of 4 Å. The closest distance between the gold atoms in this structure is **[NET DEC. 2013]**

(a) 2.828 Å
(b) 2.0 Å
(c) 2.309 Å
(d) 5.656 Å

Q.5 Polonium is the only metal known to exist in a simple cubic lattice form. The density of polonium at 0°C is measured to be 10.00 g/cm^3.

The atomic radius of polonium would then be (assume the mass of a polonium atom 2.7 x 10 g). **[NET JUNE 2017]**

(a) 1.1 Å
(b) 1.9 Å
(c) 1.5 Å
(d) 2.3 Å

Q.6 P, F and I represent primitive, face-centered and body-centered lattices, respectively. The lattice types of NaCl and CsCl, respectively, are **[GATE 2017]**

(a) F and I
(b) F and P
(c) I and P
(d) P and I

Q.7 If all the lattice points of an FCC structure are occupied by uniform hard spheres that touch each other, the fraction of volume occupied is **[NET JUNE 2018]**

(a) $\frac{\pi\sqrt{2}}{6}$

(b) $\frac{\pi\sqrt{3}}{6}$

(c) $\frac{\pi}{6}$

(d) $\frac{2\pi}{6}$

Q.8 Nearest neighbor distance in a crystal system of side length a is $\frac{a}{\sqrt{2}}$ in **[NET JUNE 2019]**

(a) Face-centered cube
(b) Body-centered cube
(c) Trigonal primitive
(d) Primitive cube

Q.9 CaTiO3 has a perovskite crystal structure. The coordination number of titanium in CaTiO3 is: **[GATE 2008]**

(a) 9
(b) 6

(c) 3
(d) 12

Q.10 The coordination number of the barium ($Ba+2$) ions in barium fluoride is 8. The coordination number of the fluoride ($F-$) ions is: **[GATE 2009]**

(a) 8
(b) 4
(c) 1
(d) 2

Q.11 The structure obtained when all the tetrahedral holes are occupied in an FCC structure is of the type **[NET JUNE 2012]**

(a) NaCl
(b) CsCl
(c) CaF2
(d) ZnS

Q.12 Coordination number of "C" in Be2C3 whose structure is correlated with that of CaF2, is: **[NET DEC. 2012]**

(a) 2
(b) 4
(c) 6
(d) 8

Q.13 The structure of CaB6 is close to that of **[NET JUNE 2018]**

(a) Cesium chloride
(b) Nickel arsenide
(c) Rock salt
(d) Zinc blende

Q.14 The INCORRECT statement about the solid-state structure of CsCl and CaF2 is **[GATE 2019]**

(a) Cation in both sides exhibit coordination number 8
(b) CsCl has BCC type structure and CaF2 has cubic close pack structure
(c) Radius ratio for Cs/Cl and Ca/F is 0.93 and 0.73, respectively
(d) Both exhibit close pack structure

Q.15 The interplanar distance (Å) for a (100) plane in a cubic structure with the lattice parameter of 4Å is: **[NET JUNE 2013]**

(a) 1
(b) 2
(c) 4
(d) 8

Q.16 The interplanar spacing of (110) planes in a cubic unit cell with lattice parameter a 4.242 Å is **[NET JUNE 2014]**

(a) 3 Å
(b) 6 Å
(c) 7.35 Å
(d) 2.45 Å

Q.17 A solid consisting of only X atoms has a close-packed structure with X-X distance of 160 pm. Assuming it to be a closed-packed structure of hard spheres with radius equal to half of the X-X bond length, the number of atoms in 1 cm3 would be **[NET DEC. 2014]**

(a) 6.023 x 1027
(b) 3.45 x 1023
(c) 6.02 × 1021
(d) 3.8 x 1021

Q.18 The distance between two successive (110) planes in a simple cubic lattice with lattice parameter "*a*" is **[GATE 2015]**

(a) $\sqrt{2}\,a$
(b) $\sqrt{3}\,a$
(c) $2\sqrt{2}\,a$
(d) $\frac{a}{\sqrt{2}}$

Q.19 For a simple cubic crystal lattice, the angle between the [201] plane and the xy plane is **[NET DEC. 2015]**

(a) less than 30°
(b) between 30° and 45°
(c) between 45° and 60°
(d) greater than 60°

Q.20 In a cubic crystal, the plane [100] is equally inclined to the planes **[NET JUNE 2016]**

(a) [010] and [011]
(b) [010] and [110]
(c) [001] and [101]
(d) [110] and [011]

Q.21 For a simple cubic lattice, the ratio between the unit cell length and the separation of two adjacent parallel crystal planes can NOT have a value of **[NET JUNE 2016]**

(a) $5^{1/2}$
(b) $7^{1/2}$
(c) $11^{1/2}$
(d) $13^{1/2}$

Q.23 Metallic silver crystallizes in face-centered-cubic lattice structure with a unit cell of length 40 mm. The first-order diffraction angle of X-ray beam from (210) plane of silver is 30°. The wavelength of X-ray used is close to **[NET JUNE 2017]**

(a) 11 nm
(b) 18 nm
(c) 25 nm
(d) 32 nm

Q.24 The separation of the (123) planes of an orthorhombic unit cell is 3.12 nm. The separation of (246) and (369) planes are, respectively **[NET DEC. 2017]**

(a) 1.56 nm and 1.04 nm
(b) 1.04 nm and 1.56 nm
(c) 3.12 nm and 1.50 nm
(d) 1.04 nm and 3.12 nm

Q.25 A metal crystallizes with cubic close-packed structure. The $\sin^2\theta$ values of Bragg reflections of Miller Planes (200) and (111) are 0.18 and 0.14, respectively. The unit cell length is **[NET DEC. 2018]**

(a) $\dfrac{\lambda}{2}$

(b) $\dfrac{\lambda}{0.2}$

(c) $\frac{\lambda}{0.4}$
(d) $0.4\,\lambda$

Q.27 In the ccp packing, the number of lattice points per unit area in the planes is in the order **[NET DEC. 2020]**

(a) $(100) > (110) > (111)$
(b) $(100) > (111) > (110)$
(c) $(111) > (100) > (110)$
(d) $(111) > (110) > (100)$

Q.28 The solid-state structures of the principal allotropes of elemental boron are made up of which of the following structural units **[DU-PhD-2018]**

(a) B4 tetrahedra
(b) B6 octahedra
(c) B8 cubes
(d) B12 icosahedra

Q.29 The number of Zn2+ ions and S2- ions are in the ZnS sphalerite unit cell **[DU-PhD-2019]**

(a) 2.4
(b) 8,8
(c) 1,1
(d) 4,4

Q.30 In a cubic crystal, the (111) and (222) reflections are observed, but not the (001) reflection. The Bravais lattice is **[NET DEC. 2018]**

(a) Body-centered cubic
(b) Face-centered cubic
(c) Simple cubic
(d) Side-centered cubic

Q.31 In a 1 g diamond crystal, a scientist wants to replace a few of the carbon atoms with nitrogen atoms. Which of the following statements best describes the resulting material? **[TIFR 2016]**

(a) This is not possible as nitrogen has a coordination number of 3 and carbon has a coordination number of 4.
(b) The resulting material will NOT have a cubic diamond structure.

(c) The resulting material will be a defective diamond as it will be p-doped.
(d) The resulting material can be characterized as n-doped diamond.

Q.32 Solid inorganic materials are commonly classified as conductors, semiconductors, or insulators as determined by their band gap. Carbon has many allotropes, such as diamond, graphite, carbon nanotubes, and graphene with distinct band gaps. Considering the properties of diamond, what should be its band gap: **[TIFR 2017]**

(a) 7.0 eV
(b) 1.0 eV
(c) 0.01 eV
(d) 0 eV

Q.33 Which of the following is a n-type semiconductor? [DU-PhD-2018]

(a) Silicon carbide
(b) Silicon
(c) Arsenic-doped silicon
(d) Gallium-doped silicon

Q.34 The diffraction of barium with X-ray radiation of wavelength 2.29 Å gives a first-order reflection at 30°. The distance between the diffracted planes is **[JNU-PhD-2018]**

(a) 2.29 Å
(b) 1.145 Å
(c) 0.709 Å
(d) 1.541 Å

Q.35 "Doping" and "Alloying" are two common terms used to describe the presence of two different elements in one material. Which of the following statements is false? **[TIFR 2017]**

(a) Doping is the process of intentionally introducing impurities into a semiconductor in order to change its electrical properties.
(b) Doping means making one solid solution.
(c) Alloying is a homogeneous mixture of two or more elements, at least one of which is a metal.
(d) In an alloy, the combined mixture has different properties from those of its component elements.

Q.36 In a face-center cubic (FCC) type of crystal lattice, the number of atoms belonging exclusively to each unit cell within the lattice is/are **[DU-PhD-2018]**

(a) 2
(b) 1
(c) 3
(d) 4

Q.37 X-rays of wavelength 1.54 Å are diffracted from a crystal at angle of 14.17°. Calculate the distance between layers in the crystal assuming first-order diffraction **[JNU-PhD-2017]**

(a) 3.1450 Å
(b) 0.3145 Å
(c) 31.4500 Å
(d) 314.5000 Å

Q.38 When crystals of sodium chloride are heated in the presence of sodium vapor, they turn yellow. This is due to the formation of **[GATE 2016]**

(a) Schottky defects
(b) Frenkel defects
(c) F-centers
(d) H-centers

Q.39 In which of the following crystals, alternate tetrahedral voids are occupied? **[JNU-PhD-2018]**

(a) NaCl
(b) ZnS
(c) CaF_2
(d) Na_2O

Q.40 When Si is doped with a Group V element. **[NET JUNE 2014]**

(a) donor levels are created close to the valence band
(b) donor levels are created close to the conduction band
(c) acceptor levels are created close to the valence bond
(d) acceptor levels are created close to the conduction band

Q.41 Zinc solid has hcp (hexagonal close-packed) structure. In a unit cell of Zinc solid, the Zinc atoms occupy **[TIFR 2017]**

(a) 74% of volume of unit cell
(b) 80% of volume of unit cell

(c) 68% of volume of unit cell
(d) 90% of volume of unit cell

Q.42 Which one of the following statements is Incorrect? **[GATE 2015]**

(a) Frenkel defect is a cation vacancy and a cation interstitial.
(b) Frenkel defect is an anion vacancy and a cation interstitial.
(c) Density of a solid remains unchanged in the case of Frenkel defects.
(d) Density of a solid decrease in the case of Schottky defects.

Q.43 When Frenkel defects are created in an otherwise perfect ionic crystal, the density of the ionic crystal **[NET DEC. 2013]**

(a) Increases
(b) Decreases
(c) Remains same
(d) Oscillates with the number of defects

Q.44 Consider a n-type semiconductor whose Ev = 0, Ec = 2.0 eV and Ed = 1.98 eV. The correct statement among the following is **[NET DEC. 2011]**

(a) Ef = l eV and is independent of T
(b) Ef = 1.99 eV and remains independent of T
(c) Ef = 1.99 eV and increases towards 2.0 eV with increase of T
(d) Ef = 1.99 eV and decreases with increase of T

Q.45 The wavelength and the spectral region for a single electron transfer across the band gap in a semiconductor (Ex = 1.98 x 10-19) are [h = 6.626 x 10-34 J.sec, c = 3 x 108 ms-1] **[NET JUNE 2012]**

(a) 1000 nm, UV
(b) 1000 nm, IR
(c) 500 nm, visible
(d) 500 nm, FAR IR

Q.46 X-ray diffraction does not give any structural information for **[NET DEC. 2014]**

(a) Metallic solids
(b) Ionic solids
(c) Molecular solids
(d) Amorphous solids

Q.47 Band theory predicts that magnesium is an insulator. However, in practice, it acts as a conductor due to **[GATE 2010]**

(a) Presence of filled 3s orbital
(b) Overlap of filled 2p and filled 3s orbital
(c) Overlap of filled 3s and empty 3p orbital
(d) Presence of unfilled 3p orbital

Q.48 The electrical conductivity of a metal [GATE 2015]

(a) Increases with increasing temperature
(b) Decreases with increasing temperature
(c) Is independent of temperature
(d) Shows oscillatory behavior with temperature

Q.49 Which of the following statements is/are not correct?

(1) The phase diagram of a substance shows the regions of pressure and temperature at which the various phases are kinetically stable.
(2) The freezing point when the pressure is 1 bar is called standard freezing point.
(3) The normal and standard freezing points are negligibly different for most of the cases.
(4) Select the correct answer using the code given below
(a) 1 only (b) 2 and 3 (c) 1 and 3 (d) 2 only

Q.50 What will be the number of phases, components, and degrees of freedom respectively of the azeotropic mixture of water and ethanol at 10 atm at equilibrium?

(a) 2, 2 and 2
(b) 1, 2 and 3
(c) 2, 2 and 0
(d) 2, 3 and 1

Q.51 For a pure substance, the triple point in a phase diagram has

(a) One degree of freedom
(b) Two degree of freedom
(c) Three degree of freedom
(d) Zero degree of freedom

Q.52 Sulfur can exist in four phases. The possible number of triple points is

(a) 1
(b) 2
(c) 3
(d) 4

Q.53 In a system, when the chemical potential of each component is the same for all the phases, the equilibrium is said to be

(a) Metastable equilibrium
(b) Thermal equilibrium
(c) Composition equilibrium
(d) Mechanical equilibrium

Q.54 The triple point for water is

(a) Unique
(b) Depends on P but is independent of T
(c) Depends on T but is independent of P
(d) Depends on both P and T

Q.55 Solid CO2 is called "dry ice" because

(a) At 25oC and 1 atm, only solid and vapor phases of CO2 are in equilibrium
(b) The critical temperature of CO2 is above 25oC
(c) The boiling point of liquid CO2 is above 100oC
(d) The melting point of solid CO2 is 0oC

Q.56 An aqueous mixed solution of NaCl and HCl is exactly neutralized by an aqueous NaOH solution. The number of components in the final mixture is **[NET DEC. 2012]**

(a) 1
(b) 2
(c) 3
(d) 4

Q.57 The lowest pressure at which the liquid phase of a pure substance can exist is known as **[NET DEC. 2012]**

(a) Critical point pressure
(b) Super-incumbent pressure
(c) Triple-point pressure
(d) Saturation vapor pressure

Q.58 The number of phases, components and degree of freedom, when Ar is added to an equilibrium mixture of NO, O2, and NO2 in gas phase are, respectively **[NET DEC. 2013]**

(a) 1, 3, 5

(b) 1, 4, 5
(c) 1, 3, 4
(d) 1, 4, 4

Q.59 The number of degrees of freedom of a system consisting of solid sucrose in equilibrium with an aqueous solution of sucrose is

(a) 0
(b) 1
(c) 2
(d) 3

Q.60 Solid carbon reacts with oxygen in the presence of a catalyst to form the gaseous oxides CO and CO2. What is the number of degrees of freedom (variance) for the system once equilibrium has been attained?

(a) 0
(b) 1
(c) 2
(d) 3

Q.61 The vapor of pure substance, when cooled under a pressure less that its triple-point pressure **[NET JUNE 2012]**

(a) Liquefies
(b) Liquefies first and then solidifies
(b) Solidifies directly
(d) Remains unchanged

Q.62 Triple-point pressure of substance A, B, C, and D are 0.2, 0.5, 0.8, and 1.2 bar, respectively. The substance which sublimes under standard conditions on increasing temperature is **[NET DEC. 2016]**

(a) A
(b) B
(c) C
(d) D

Q.63 A phase transition process is always **[NET JUNE 2017]**

(a) isothermal – isoentropic
(b) isochoric – isothermal
(c) isobaric – isochoric
(d) isothermal – isobaric

Q.64 m1 and m2 are the slopes (dP/dT) of the solid–liquid equilibrium lines in the P-T phase diagrams of H2O and CO2, respectively. For P < 10 atm, the values of m1 and m2 are **[GATE 2017]**

(a) m1 > 0 and m2 > 0
(b) m1 > 0 and m2 < 0
(c) m1 < 0 and m2 < 0
(d) m1 < 0 and m2 > 0

Q.65 In the phase diagram of water, the solid–liquid boundary has a negative slope. The reason for this unusual behavior can be traced to decrease in **[NET JUNE 2018]**

(a) Density of the system on melting
(b) Volume of the system on the melting
(c) Entropy of the system on melting
(d) Enthalpy of the system on melting

Q.66 Consider a phase transition between two incompressible phases. The correct statement among the following is **[NET DEC. 2018]**

(a) The transition is independent of pressure
(b) The transition is independent of temperature
(c) The entropy of such transition is always zero
(d) The enthalpy of such transition is always non-zero

Q.67 The volume change in a certain phase transition is 2.0 ml.mol-1 at transition point. From this, we may conclude that the transition is most likely to **[NET DEC. 2018]**

(a) First-order phase transition
(b) Second-order phase transition
(c) Third-order phase transition
(d) λ phase transition

Q.68 For a cubic crystal system, the powder X-ray diffraction pattern recorded using Cu K_α source (λ = 1.54Å) shows a peak at 33.60° (2θ) for (111) plane. The lattice parameter "a" (in Å. rounded off to two decimal places) is_______________ **[GATE 2020]**

Answer Key

1.(a)	**2.(b)**	**3.(d)**	**4.(a)**
5.(c)	**6.(b)**	**7.(a)**	**8.(a)**
9.(b)	**10.(b)**	**11.(c)**	**12.(d)**
13.(a)	**14.(a)**	**15.(c)**	**16.(a)**
17.(b)	**18.(d)**	**19.(d)**	**20.(a)**
21.(b)	**23.(b)**	**24.(a)**	**25.(c)**
27.(c)	**28.(d)**	**29.(d)**	**30.(b)**
31.(d)	**32.(a)**	**33.(c)**	**34.(a)**
35.(b)	**36.(d)**	**37.(a)**	**38.(c)**
39.(b)	**40.(b)**	**41.(a)**	**42.(b)**
43.(c)	**44.(d)**	**45.(b)**	**46.(d)**
47.(b)	**48.(b)**	**49.(a)**	**50.(c)**
51.(d)	**52.(d)**	**53.(c)**	**54.(a)**
55.(a)	**56.(b)**	**57.(a)**	**58.(c)**
59.(c)	**60.(d)**	**61.(c)**	**62.(d)**
63.(d)	**64.(d)**	**65.(b)**	**66.(a)**
67.(a)	**68. (4.62)**		

Index

D

E

N

O

P